Peptides

Biology and Chemistry

Peptides

Biology and Chemistry

Proceedings of the 1996 Chinese Peptide Symposium
July 21-25, 1996, Chengdu, China

Edited by

Xiao-Jie Xu
Department of Chemistry
Peking University
Beijing 100871, China

Yun-Hua Ye
Department of Chemistry
Peking University
Beijing 100871, China

and

James P. Tam
Department of Microbiology and Immunology
Vanderbilt University
Nashville, TN 37232, U.S.A.

Springer Science+Business Media, B.V. 1998

A C.I.P. Catalogue record for this book is available from the Library of Congress.

ISBN 978-94-017-4108-8 ISBN 978-0-306-46859-9 (eBook)
DOI 10.1007/978-0-306-46859-9

Printed on acid-free paper

Preface

The fourth Chinese Peptide Symposium, hosted by Peking University, was held at Chengdu, China on July 21-25, 1996 with 164 participants, including 45 scientists from abroad, representing 12 countries. The four-day conference was both intense and spiritually rewarding. Our goal for CPS-96 was to provide a forum for the exchange of knowledge, cooperation and friendship between the international and Chinese scientific communities, and we believe this goal was met.

The symposium consisted of 10 sessions with 55 oral and 78 poster presentations, including synthetic methods, molecular diversity and peptide libraries, structure and conformation of peptides and proteins, bioactive peptides, peptide immunology, De Novo design and synthesis of proteins and peptides, ligand-receptor interactions, the chemistry-biology-interface and challenging problems in peptides.

The enthusiastic cooperation and excellent contributions were gratifying and the active response of the invited speakers contributed to the success of the symposium. The presentations were of excellent caliber and represented the most current and significant aspects of peptide science.

Dr. James P. Tam and Dr. Jie-Cheng Xu were the recipients of 'The Cathay Award' sponsored by the H. H. Liu Education Foundation, offered for their seminal contributions in peptide science and the Chinese Peptide Symposium. Four outstanding young scientists were selected by the organizing committee to receive awards sponsored by Haikou Nanhai Pharmaceutical Industry Co. Ltd. (Zhong He Group).

It is our pleasure to acknowledge Nobel Laureate Professor Bruce R. Merrifield for his kindness in supporting and serving as the chairman of the Awarding Committee for the 'Cathay Award', and all members of the program and organizing committee for their generous contributions to the successful symposium.

Dr. John R. Martin and Kluwer Academic Publishers kindly agreed to publish the CPS-96 proceedings, and we thank them for their contributions and support of the symposium.

We greatly appreciate the generous financial assistance of the sponsors and donors. On behalf of the organizing committee, we thank the administrative assistance by Peking University and Sichuan University as well as Vanderbilt University.

Finally, we are grateful to our colleagues, especially Professor Yan-Ling Song, Professor Gui-Lin Tian, Dr. Jia-Xi Xu, Dr. Zhen-Wei Miao and Ms. Vicki Gates Bryant, who have devoted much time and effort to ensure the success of CPS-96.

Xiao-Jie Xu
Yun-Hua Ye
James P. Tam

Chinese Peptide Symposium - 1996

July 21-25, 1996, Chengdu, China
Peking University
Sichuan University

Chairpersons

Dr. Xiao-Jie Xu, *Peking University, Beijing, China*
Dr. Yun-Hua Ye, *Peking University, Beijing, China*

Program Committee

Yu-Cang Du, Chairman, *Shanghai Institute of Biochemistry, Shanghai, China*
James P. Tam, *Vanderbilt University, U.S.A.*
Xiao-Jie Xu, *Peking University, Beijing, China*
Meng-Shen Cai, *Beijing Medical University, Beijing, China*
Roger Epton, *Mayflower Worldwide Ltd., U. K.*
Yue-Ting Gong, *Shanghai Institute of Biochemistry, Shanghai, China*
Victor J. Hruby, *University of Arizona, U.S.A.*
Chong-Xi Li, *Peking University, Beijing, China*
Jean Martinez, *CNRS Faculte de Pharmacie, France*
Jean E. Rivier, *The Salk Institute, U.S.A.*
Peter W. Schiller, *Clinical Research Institute of Montreal, Canada*
Yasutsugu Shimonishi, *Osaka University, Japan*
Akinori Suzuki, *University of Tokyo, Japan*
John D. Wade, *University of Melbourne, Australia*
Jie-Cheng Xu, *Shanghai Institute of Organic Chemistry, Shanghai, China*

Organizing Committee

Yun-Hua Ye, Chairman, *Peking University, Beijing, China*
Gui-Shen Lu, Vice-chairman, *Institute of Materia Medica, Beijing, China*
James P. Tam, Vice-chairman, *Vanderbilt University, U.S.A.*
Saburo Aimoto, *Osaka University, Japan*
Shu-Hua Chen, *Sichuan University, Chengdu, China*

Zhen-Kai Ding, *Chinese Institute of Military Medical Science, Beijing, China*
Xiao-Yu Hu, *Lanzhou University, Lanzhou, China*
Wei Li, *Jilin University, Changchun, China*
Robert Liu, *H. H. Liu Education Foundation, U.S.A.*
Bing-Gen Ru, *Peking University, Beijing, China*
Arnold Satterthwait, *The Scripps Research Institute, U.S.A.*
Shao-Bo Xiao, *State Family Planning Commission, Beijing, China*

Awarding Committee of Cathay Award

Bruce Merrifield, Chairman, *The Rockefeller University, U.S.A.*
Victor J. Hruby, *University of Arizona, U.S.A.*
Peter W. Schiller, *Clinical Research Institute of Montreal, Canada*
Yu-Cang Du, *Shanghai Institute of Biochemistry, Shanghai, China*
Gui-Shen Lu, *Institute of Materia Medica, Beijing, China*
Zhen-Kai Ding, *Chinese Institute of Military Medical Science, China*
Xiao-Jie Xu, *Peking University, Beijing, China*

Major Sponsors

H. H. Liu Education Foundation
National Natural Science Foundation of China

Sponsors

ASTA MEDICA
Penninsula Laboratories, Inc.
Pharmacia Biotech
Beaufour Ipsen International China Office
American Peptide Company, Inc.
Eli Lilly and Company
Torrey Pines Institute for Molecular Studies
Amgen, Inc.
Isochem
Penta Biotech, Inc.
Warner-Lambert Company
Beijing Xingda Science System Co.

Donors

Chiron Corporation
Waters Corporation
World Ways Bio Electronics Co.

Abbreviations

Abbreviations used in the proceedings volume are defined below:

2D-NMR	two dimensional NMR
AAA	amino acid analysis
ACE	angiotensin-converting enzyme
ACh	acetylcholine
Acm	acetamidomethyl
ADP	adenosine diphosphate
Aib	aminoisobutyric acid
AIDS	acquired immune deficiency syndrome
AM	adrenomedullin
AMD	actinomycin D
AMP	adenosine monophosphate
Aoq	α-amino-β-[4-(1,2-dihydro-2-oxo quinolinyl)]propionic acid
APC	antigen presenting cell
ATP	adenosine triphosphate
Atq	2-amino-3-(1,2-dihydro-2-thio-quinolinyl) propionic acid
AVP	arginine-8-vasopressin
Boc	*tert*-butyloxycarbonyl
BSA	bovine serum albumin
Bu, Bu^t	butyl
Bzl	benzyl
bZIP	basic region-leucine zipper
CaM	calmodulin
CaMKII	Ca^{2+}/CaM-dependent protein kinase
Cbz	carbobenzoxy; benzyloxycarbonyl
CCK	cholecystokinin
CD	circular dichroism; complement domain
CE	capillary electro-phoresis
CGRP	calcitonin gene-related peptide
Cha	cyclohexylamine
CPS	Chinese peptide symposium
CT	carboxy terminus; calcitonin; chymotrypsin; cholera toxin
CTL	cytotoxic T-lymphocytes
D	diversity (as with Ig or TCR genes)
DAPaa	*N*-(O,O-dialkyl) phospho amino acid
DCC	dicyclohexyl-carbodiimide
DCM	dichloromethane
DDSi	dichlorodimethylsilane
DEPBT	3-diethoxyphosphoryloxy-1,2,3-benzotriazin-4(3H)-one
DIEA	diisopropylethylamine
DIPEA	diisopropylethylamine
DIPP	*N*-(O,O-diisopropyl)-phosphoryl
DMF	dimethylformamide
DMSO	dimethyl sulfoxide
DOPC	dioleoyl-sn-glycero-phosphocholine
DTT	dithiothreitol
EDC	1-(3-dimethylaminopropyl)-3-ethylcarbodiimide hydrochloride
EDTA	ethylenediamine-tetraacetic acid
ELISA	enzyme-linked immunosorbent assay

ERK	extracellular signal-regulated kinases
FAB-MS	fast atom bombardment mass spectrometry
FD	flexible docking
Fmoc	9-fluorenylmethoxycarbonyl
FSH	follicle stimulating hormone
Ftase	farnesyl transferase
FTIR	Fourier transform infrared
GH	growth hormone
GHR	growth hormone receptor
GPCR	G-protein coupled receptor
GSTs	glutathione s-transferases
HBTU	*O*-benzotriazolyl-*N,N,N',N'*-tetramethyluronium hexafluorophosphate
hCG	human chorionic gonadotropin
hCG-PP	hCG-induced progesterone production
HCV	hepatitis C virus
HEV	hepatitis E virus
HIV	human immuno-deficiency virus
HLA	human leukocyte antigen
HOBt	*N*-hydroxybenzotriazole
HOOBt	*N*-hydroxyoxodihydrobenzotriazine
HOSu	*N*-hydroxysuccinimide
HPLC	high performance liquid chromatography
Ig	immunoglobulin
IL	interleukin
IPTG	isopropyl-D-thiogalactoside
IR	infrared; insulin receptor
LH	luteinizing hormone
LHRH	luteinizing hormone releasing hormone
LTP	long-term potentiation
MAb	monoclonal antibody
MALD-MS	matrix-assisted laser desorption mass spectrometry
MAP	multiple antigen peptide; mean arterial pressure
MAPK	mitogen-activated protein kinase
MBHA	methylbenzhydrylamine
MDP	muramyl dipeptide
ME	mercaptoethanol
MHC	major histocompatibility complex
MM	molecular mechanics
MP	mastoparan; melting point
MS	mass spectrometry
MT	metallothionein
MTT	3-(2-yl)- 2,5-diphenyl-tetrazolium bromide
MVD	mouse vas deferens
NGF	nerve growth factor
Nle	norleucine
NMDA	*N*-methyl-D-aspartate
NMDA-R	NMDA-receptor
NMM	*N*-methylmorpholine
NMR	nuclear magnetic resonance
OEt	O-ethyl ester
OMe	O-methyl ester
OSu	O-succinimide ester
PACAP	pituitary adenylatecyclase

activating polypeptide
PBMC peripheral blood mononuclear cell
PBS phosphate buffered saline
PCR polymerase chain reaction
PDB phorbol 12,13-dibutyrate;
PDB protein data bank
PDGF platelet derived growth factor
PDI protein disulfide isomerase
PEG polyethylene glycol
Pfp pentafluorophenyl ester
PLGA copolymer of lactic acid-co-glycolic acid
PSPP papaver somniferum pollen peptide
PTH parathyroid hormone
PTX pertussis toxin
Py pyridine

RBD rigid-body docking
RP-HPLC reversed-phase high performance liquid chromatography
RTK receptor tyrosine kinase

SAg superantigen
SD soft docking
SDS sodium dodecyl sulfate
SFD semiflexible docking
Snm *S*-(N'-methyl-N-phenylcarbamoyl) sulfenyl
SP substance P
SPPS solid phase peptide synthesis

TASP template assembled synthetic protein
TCEP tris(2-carboxyethyl)-phosphine
TCR T lymphocyte antigen receptor
TEA triethylamine
TFA trifluoroacetic acid
TFE trifluoroethanol
TFMSA trifluoromethane-sulfonic acid
THF tetrahydrofuran
Tic 1,2,3,4-tetrahydro quinoline-3-carboxylic acid
TIPP H-Tyr-Tic-Phe-Phe-OH
TIS triisopropyl silane
TLC thin layer chromatography
TLN thermolysin
Tmob *S*-2,4,6-trimethoxybenzyl
TMSBr trimethylsilyl bromide
TNF tumor necrosis factor
TOF-MS Time of flight mass spectra
Tris tris(hydroxymeth-yl) aminomethane
TSST Toxic shock syndrome toxin

UNCA N-urethane-protected carboxyanhydride
UV ultraviolet
VIP vasoactive intestinal peptide

Z carbobenzoxy

Contents

Session II: Molecular Diversity-Peptide Libraries

Session III Conformation of Peptides and Proteins

Session IV: Structure and Design

Session V: Bioactive Peptides

Session I
Approach and Methods

Chairs: Jean Martinez
CNRS Faculté de Pharmacie
France

Kit S. Lam
Arizona Cancer Center
University of Arizona
Tucson, Arizona, U.S.A.

Xiao-Yu Hu
Lanzhou University
Lanzhou, China

and

Shao-Bo Xiao
State Family Planning Commission
Beijing, China

Cyclic peptides from unprotected precursors through ring-chain tautomerism

James P. Tam, Lian-Shan Zhang, Paolo Botti and Yi-An Lu
Department of Microbiology and Immunology, Vanderbilt University, A 5119 MCN, Nashville, TN 37232, U.S.A.

Introduction

In 1993, our laboratory introduced a novel approach of blockwise ligation through amide bonds for peptide synthesis using unprotected peptide segments under aqueous conditions [1]. This approach uses a pair of mutually reactive groups to achieve selective amide bond formation between a specific α-amine on one peptide and an α-acyl moiety on another peptide with both peptides containing free α- and ϵ-amines. Because of its exclusive chemoselectivity, we have referred to this approach as the orthogonal coupling strategy [1-5]. By exploiting different pairs of reactive groups, several methods based on this strategy have been developed and applied successfully to the synthesis of proteins [5, 6]. In addition, we have found that the orthogonal coupling strategy is well suited for preparing cyclic peptides [7,8] when two reactive ends are present on a single chain, largely due to unprotected peptide segments undergoing the entropy-favored ring-chain tautomerization.

Ring-chain tautomerization is a novel concept and has the attendant advantage of eliminating the requirement for high dilution in the macrocyclization of unprotected peptides. In this paper, we describe the concept of ring-chain tautomerization in combination with the orthogonal ligation strategy for intramolecular cyclization of unprotected peptides to form end-to-end cyclic peptides using two different orthogonal coupling methods.

Results and Discussion

Conceptual framework

Central to the orthogonal coupling strategy is the amide formation through entropic activation [9]. It differs both in concept and mechanism from the conventional, enthalpic coupling methods and has three key features: (1) the use of a weakly activated α-acyl moiety, generally unreactive to amino groups, but reactive with an amino group proximally linked with another strong nucleophile; (2) amide bond formation proceeds through at least two discrete chemical steps, one of which involves an acyl rearrangement; and (3) the initial attack on the α-acyl moiety does not involve the α-amine but the proximally linked nucleophile and is often reversible under aqueous conditions. These features are absent in the conventional coupling strategies with strongly activated α-acyl moieties which lead to random acylations, necessitate the use of multi-tiered protecting groups, and increase the danger of racemization. The use of a weakly activated α-acyl moiety that equilibrates among different nucleophiles also provides the foundation for the ring-chain tautomerization to minimize unwanted oligomerization.

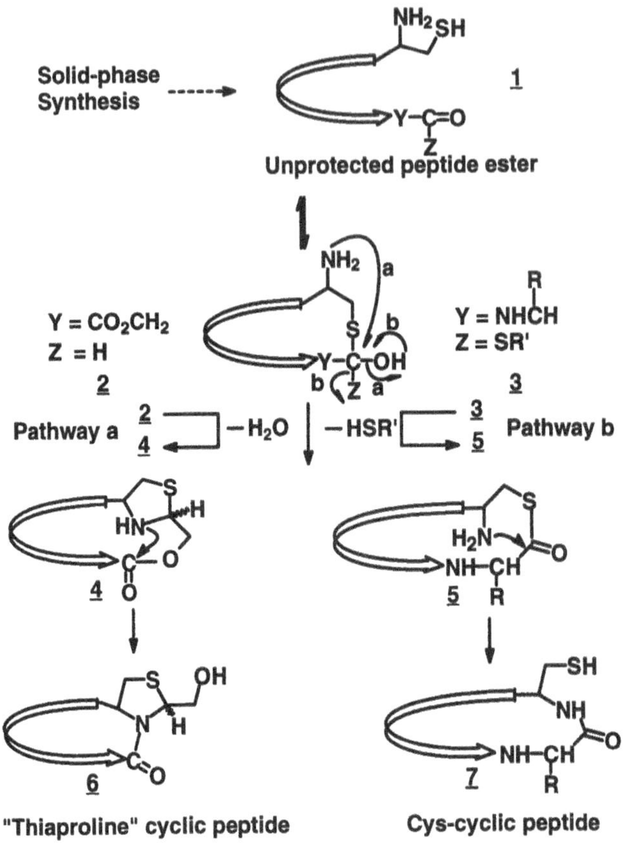

Fig. 1. A general scheme for the preparation of cyclic peptides utilizing ring-chain tautomerization. R' = -$CH_2CH_2CONH_2$ or -CH_2CH_2CONH-CH_2COOH.

In ring-chain tautomerization (Fig. 1), an open chain system **1** containing a 1,2-aminothiol at one end and a polar multiple bond at the other end is transformed into a cyclic intermediate **2** or **3** through reversible intramolecular addition of the thiol group to the polar multiple bond. Either a water, thiol or alcohol molecule is then eliminated to yield a stable nonisomeric heterocycle **4** or **5**. By positioning the acyl group and the amine moiety in close proximity, intramolecular acyl migration can be effected to give an amide bond in compound **6** or **7**. This principle is successfully demonstrated in two different types of open chain systems. In the first type, an open chain system of an amino-aldehyde peptide containing an N-terminal cysteine and a C-terminal glycolaldehyde ester (Fig. 1, Pathway a) was used. The initial thiazolidine ring formation is an intermediate step that leads to a cyclic lactam as the final product. In the second type, an open chain of an N-terminal cysteinyl peptide thioester can be considered as a variation of the first type. The initial step is a reversible transthioesterification. The cyclic thiolactone intermediate will subsequently rearrange into a cyclic lactam **7** through a 5-membered ring S to N-acyl transfer (Fig. 1, Pathway b).

Thiaproline cyclization through cysteinyl peptide aldehyde

This method of cyclization provides an end-to-end cyclic peptide containing a thiazolidine residue in the ring system. Thiazolidine is a thiaproline analog and is also the product of the intramolecular ligation between an N-terminal 1,2-aminothiol of Cys and a carboxyl glycolaldehyde in an unprotected peptide precursor (Fig. 2a). For preparing an aldehyde at the C-terminus in a precursor form, we used a glyceric ester derived from an acetal peptide-resin [7]. The peptide precursors were assembled on an acetal resin using Fmoc chemistry, and the peptides were liberated by TFA which contained a masked aldehyde as the diol. Conversion of the diol to the glycolaldehyde was effected by oxidation with $NaIO_4$ for 30 min at pH 5. To protect the cysteinyl thiol during the oxidation, the t-butylsulfenyl group was used, and the thiol was regenerated by treatment with trialkylphosphine (TECP) at pH 5.5. Concomitant cyclization also occurred to give a cyclic peptide with a thiazolidine linkage within 1-6 h for cyclic peptides containing 5, 6,12 and 26 amino acid residues. The intramolecular O to N-acyl transfer to the cyclic peptides with an amide backbone occurred after the pH of the buffer was adjusted to 5.9 but required more than 20 h for completion at 22°C. Increasing the temperature to 52°C raised the reaction rate 3 to 5 fold.

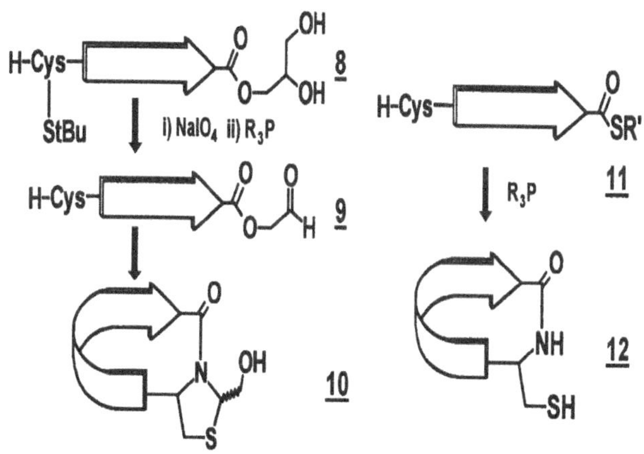

*Fig. 2. (a) Thiaproline cyclization (left). Peptide sequence: **8a** CGRAG; **8b** CGRAFG; **8c** CGRAFVTIG; **8d** CGRAFVTICKIG; **8e** CNNNTRKRIRIQRGPGPGRAFVTIGKIG. (b) Cysteinyl cyclization (right). Peptide sequence: **11a** CGGFL; **11b** CYGGFL; **11c** CKYGGFL; **11d** CKAYGGFL; **11e** CAVSEIQFMHNLGK; **11f** CSNLSTCVLGKLSQEL.* $R = -CH_2CH_2COOH$. $R' = -CH_2CH_2CONH_2$ *or* $-CH_2CH_2CONHCH_2COOH$.

Cysteinyl cyclization through transthioesterification

This method of cyclization produces an end-to-end cyclic peptide containing a Cys which is also the ligation site (Fig. 2b). It requires a free peptide containing an N-terminal cysteinyl and a carboxyl thioester. Peptide thioesters were prepared directly by solid-phase peptide synthesis using Boc chemistry on either a MBHA resin or Gly-Pam resin prepared according to Aimoto et al. [10]. Peptides ranging from 5 to 16 amino acid residues were used for cyclizations directly after cleavage from the resin support because the crude unprotected cysteinyl peptide thioesters were found to be sufficiently pure (>87%). Cyclizations were performed at or around pH 7 in 0.2M phosphate buffer. Tris(carboxyethyl)phosphine (TCEP) was added to prevent disulfide formation and to accelerate the desired reaction. The cyclization which included both steps of transthioesterification and the spontaneous S to N-acyl shift occurred cleanly and was completed within 4 h as shown by RP-HPLC in yields ranging from 78 to 92%.

Evidence to support ring-chain tautomerization

The concentration-independent nature of cyclization and the lack of polymeric products in the RP-HPLC traces were used as evidence to support the ring-chain tautomerization in our cyclization process. In this study, cyclizations were run in moderate to high concentrations at millimolar levels without affecting the yield. For example, cyclizations via transthioesterification of **11b, 11c, 11d, 11e** and **11f** were performed in 1, 3.5 and 7 mM at pH 7.5. Oligomerization of the peptide which is often the case in cyclization using protected peptide segments was not detected by RP-HPLC. In all three concentrations, more than 90% of cyclic peptides were recovered as the isolated yield after RP-HPLC purification. Similar results were also obtained from the thiazolidine cyclization of **8d,** a 12-residue peptide, at 0.6, 6 and 20 mM concentrations at pH 5.8, and no polymerization was observed. In the case of **11e,** the reaction could be carried out in a concentration of 20 mM of the peptide, which was > 200 fold higher than that of the conventional approach using active ester and high dilution. Thus, our results support the concentration-independent cyclization and conform with the expected equilibrium behavior of ring-chain tautomerism.

Isolable stable intermediates during amide bond formation

Since the amide bond formation in our approach involves discrete chemical steps and is significantly different from the conventional coupling methods, the strained cyclic peptides provide an opportunity to highlight the presence of these intermediates in our coupling strategy. In the thiazolidine cyclization, the thiazolidines prior to the acyl transfer were stable compounds at pH < 5.0 because the O to N-acyl transfer reaction passes through a highly strained tricyclic intermediate undergoing ring contraction after adjustment of the pH to 5.9. In cyclization via transthioesterification, the cyclic thiolactone is usually not observed when the reactions are performed at pH 6.5 to 7.5. This cyclic thiolactone rapidly rearranged through an S to N-acyl shift to the desired end-to-end cyclic peptide. To show the cyclization intermediate of thiolactone, we used a Cys-Gly-Gly-Phe-Leu pentapeptide thioester **11a**. In this case, the cyclic thiolactone intermediate is a 16-membered ring, and a ring contraction via an S to N-acyl shift gives a 15-membered end-to-end cyclic peptide. Because of the ring strain, the contraction via S to N-acyl shift was slow at pH 4.5, and it was possible to isolate the labile cyclic thiolactone intermediate which was confirmed by MALDI-MS (found: 499.5, calcd for $[M+Na-H]^+$: 499.5) and its susceptibility to hydroxylamine at pH 9. In our study of the intermolecular transthioesterification of two unprotected peptide segments, the S to N-acyl transfer occurs spontaneously, and thus far the thioester intermediate has not been observed. Therefore, it is interesting that we could confirm the thioester intermediate using this cyclic pentapeptide model.

Conclusion

The orthogonal coupling strategy of amide ligation using unprotected peptide segments provides a facile approach for the synthesis of both small and large peptides. For small to moderate-size peptides, we have exploited the entropy-favored ring-chain tautomerization to facilitate intramolecular cyclization with high efficiency. The importance of cyclic peptides in the drug discovery process has been widely exploited due to their conformational rigidity and resistance to degradation. Recently, cyclic peptides have been used for libraries for this purpose. Because of their importance, an efficient and conceptually novel method for their synthesis is highly desirable. Conventional methods for synthesizing cyclic peptides require the protection of selected functional groups, multitiered protection-deprotection schemes and cumbersome purification of linear protected precursors and cyclic products, and the use of organic solvents at high dilution to prevent competing intermolecular oligomerization. Thus, the development of novel cyclization methods based on ring-chain tautomerization in conjunction with the orthogonal coupling strategy through unprotected peptide segments will be useful for generation of cyclic peptide libraries.

Acknowledgments

This work was in part supported by US Public Health Service NIH Grant AI37965 and CA35644.

References

1. Liu, C.F., Shao, J., Rao, C. and Tam, J.P., In Hodges, R.S. and Smith, J.A. (Eds.) Peptides: Chemistry, Structure and Biology (Proceedings of the Thirteenth American Peptide Symposium), ESCOM, Leiden, The Netherlands, 1994, p.218.
2. Liu, C.F. and Tam, J.P., J. Am. Chem. Soc., 116 (1994) 4149.
3. Tam, J.P., Lu, Y.-A., Liu, C.F. and Shao, J., Proc. Natl. Acad. Sci. U.S.A.,92 (1995) 12485.
4. Dawson, P.E., Muir, T.W., Clark-Lewis, I. and Kent, S.B.H., Science, 266 (1994) 776.
5. Liu, C.F., Rao, C. and Tam, J.P., Tetrahedron Lett., 37 (1996) 933.
6. Liu, C.F., Rao, C. and Tam. J.P., J. Am. Chem. Soc., 118 (1996) 307.
7. Botti, P., Pallin, D.P. and Tam, J.P., J. Am. Chem. Soc., 118 (1996) 10018.
8. Zhang, L and Tam, J.P., J. Am. Chem. Soc., in press.
9. Kemp, D.S., Biopolymers, 20 (1991) 1793.
10. Hojo, H. and Aimoto, S., Bull. Chem. Soc. Jpn., 64 (1991) 111.

Synthesis of glycopeptides with Lewis antigen or sialyl-Lewis^x antigen side chains

Horst Kunz[a], Karsten von dem Bruch[a], G. Kretzschmar[b] and Ulrich Sprengard[a]
[a] *Institut für Organische Chemie, Universität Mainz, D-55099 Mainz, Germany*
[b] *Hoechst AG, Zentralforschung, D-65926 Frankfurt, Germany*

Introduction

During the past decades it has been recognized that the glycosylation pattern of glycoproteins plays a key role in processes such as the transport of proteins through the endoplasmic reticulum or membranes, the docking of bacteria and viruses onto cells, and cell-cell recognition. Glycoproteins which carry Lewis antigen saccharide side chains **1** and **2** were identified as cell surface markers in cell differentiation. For example, the Lewis^x determinant **1** is formed during embryonic development at the 8-cell stage by α-1,3-fucosylation of the I determinant as a status-specific embryonic antigen (SSEA-1). Later, it is almost completely masked by fucosylation (to give the Lewis^x antigen) or sialylation [1]. Glycoproteins with sialyl Lewis^x saccharide portions **3** are important ligands of receptors located in membranes of endothelial cells (E-selectin and P-selectin), platelets (P-selectin) and leukocytes (L-selectin). The recognition of sialyl Lewis^x ligands by selectins is obviously responsible for the early steps in cell adhesion and is of increasing interest in context with inflammatory processes. It controls the recruitment of leukocytes and their subsequent invasion into the center of inflammation [2]. In contrast to sialyl Lewis^x, the occurrence of Lewis^x antigen **1** in healthy adults is restricted to certain specific regions.

For tumors it has been shown, that the accumulation of Lewis^x antigen in the cell membranes directly corresponds to the extent of loss of cell-cell recognition. Therefore, glycoconjugates with Lewis^x antigen structure **1** are of interest as tumor-associated antigens. For glycoproteins, it seems inappropriate to rigorously separate the role of carbohydrates and peptides since both portions contain numerous polar functionalities. Model compounds for the investigation of recognition phenomena involving glycoproteins should have a defined structure in both the carbohydrate and the peptide portion.

Synthetic Lewis^x antigen glycopeptides

In addition to the general demands of glyco-peptide synthesis arising from the sensitivity of glycosidic bonds [4], the synthesis of glycopeptides with Lewis antigen and sialyl Lewis antigen saccharides includes additional difficulties: a) The α-fucoside bond is particularly sensitive to acids and can readily be cleaved by formic acid [5]. b) The stereo- and regioselective introduction of a sialic acid residue is problematic, and the sialic acid contains a carboxylic function which must be protected reversibly and differentiated from the peptide carboxylic functions.

The first problem in the synthesis of glycopeptides with complex saccharide side chains connsists of the stereo- and regioselective synthesis of the corresponding oligosaccharide. In order to provide a versatile efficient coupling of the saccharide to the amino acid or peptide component, we used the azido group as the anomeric protecting group of the

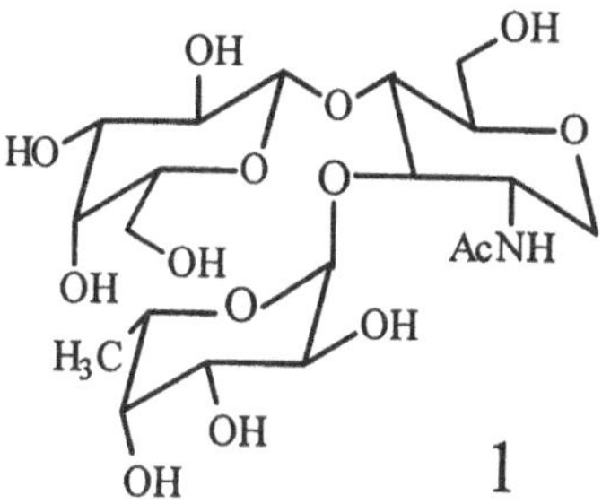

1

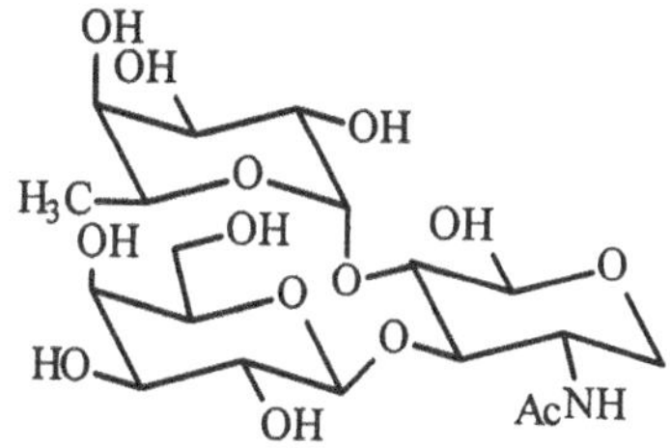

2

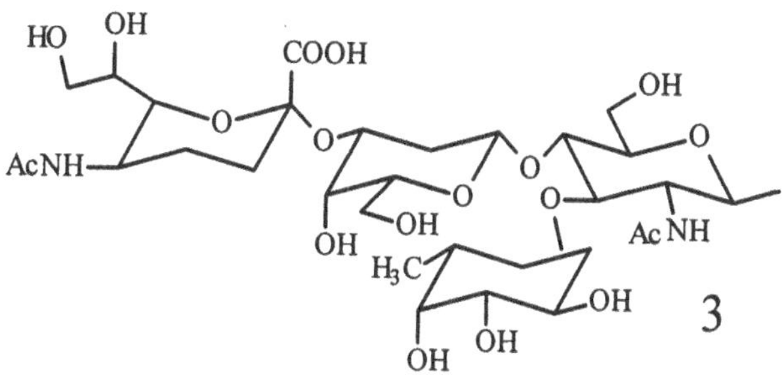

3

glucosamine portion [6]. Subsequent introduction of the β-1,4-galactosyl and α-1,3-fucosyl unit gave the protected Lewis^x trisaccharide azide **4**. In order to obtain acid-stability of the sensitive α-fucoside bond which is required for glycopeptide synthesis, the O-(4-methoxy-benzyl)ether protecting groups (Mpm) were exchanged for O-acetyl groups. Then, the azido group was hydrogenolized to give the Lewis^x trisaccharide amine **5**. Its condensation with α-allyl Boc aspartate yielded the Lewis^x asparagine conjugate **6**. The protecting group combination of **6** allows theselective and alternative removal of either the Boc group or the allyl ester. Using the selectively deblocked components **7** and **8**, the glycohexapeptide which carries two Lewis^x antigen side chains was synthesized and completely deprotected to give **9**. Coupling of **9** to bovine serum albumin (BSA) and keyhole limpet haemocyamine yielded neoglycoproteins which contain on average 28 synthetic Lewis^x glycopeptides **9** of exactly specified structure per molecule protein [7].

Sialyl Lewis^x glycopeptides

As for the construction of the Lewis^x antigen saccharide, the azido group was also used as the protecting group and precursor of the anomeric amino group during synthesis of the sialyl Lewis^x tetrasaccharide portion. In this synthesis, α-1,3-fucosylation was carried out first. After reductive opening of the benzylidene acetal, the β-1,4-galactosylation was carried out using the trichloroacetimidate of 6-O-benzyl-2,3,4-tri-O-acetyl-galactopyranose. De-O-acetylation yielded the Lewis^x trisaccharide **10**. The sialylation of **10** was achieved using the methyl thiosialoside **11** and activation with methylsulfenyl trifluoromethane sulfonate to give the sialyl Lewis^x azide **12** with high regio- and stereoselectivity [8].

For an application in glycopeptide synthesis, the methyl ester of the sialyl acid portion of **12** must be saponified, since treatment of glycopeptides with alkaline media would result

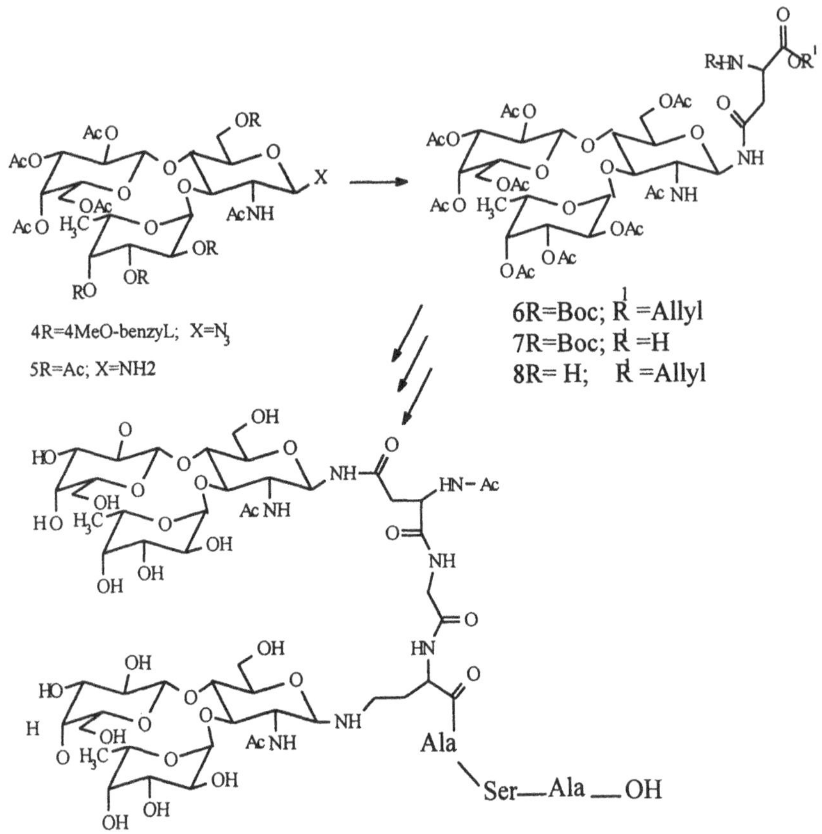

in racemization (epimerization) within the peptide and cleavage of certain glycosidic bonds.Treating a solution of saponified **12** with acidic ion-exchange resin, the sialyl Lewis^x lactone **13** was formed. We considered this lactone a useful internal protection of the sialyl carboxylic function and hydrogenolyzed the anomeric azido group to obtain the sialyl Lewis^x lactone amine **14**. Its condensation with preformed peptides, e. g. the RGD peptide **15** proved, that the lactone is a reliable internal protection of the sialic carboxylic group. The formed sialyl Lewis^x RGD peptide conjugate was subjected to complete deprotection and opening of the lactone to give **16** in high overall yield. The sialyl Lewis^x RGD peptide proved to be only a weak ligand for E-selectin but a very potent inhibitor of P-selectin showing an IC_{50} of 26 μm. This result suggests that the peptide portion efficiently modulates the recognition properties of the sialyl Lewis^x motif in cell adhesion processes and encourages us to carry out further extensive studies in this field.

Acknowledgment

This work was supported by the Deutsche Forschungsgemeinschaft (SFB 302 and SFB 519) and by the Hoechst AG.

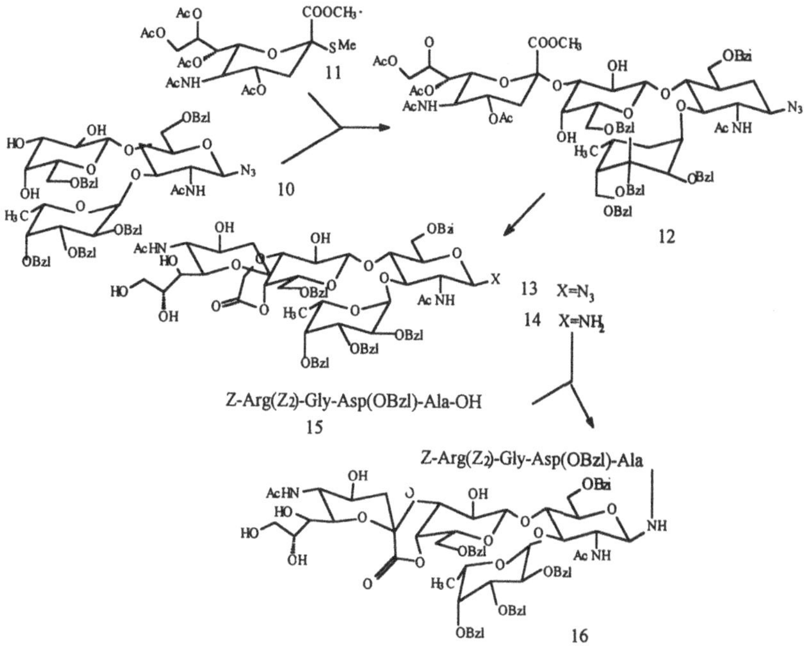

References

1. Gooi, H. C., Eur. J. Immunol. 13 (1983) 306.
2. Mulligan, M. S., Paulson, J. C., DeFrees, S., Zheng. Z.-L., Lowe, J. B. and Ward, P. A., Nature 364 (1993) 149.
3. Hounsell, E. F., Chem. Soc. Rev. 16 (1987), 161.
4. Kunz, H., Pure Appl. Chem. 65 (1993) 1223.
5. Unverzagt, C. and Kunz, H., Bioorg. Med. Chem. 2 (1994) 1189.
6. Unverzagt, C. and Kunz, H., J. prakt. Chem. 334 (1992) 570.
7. von dem Bruch, K. and Kunz, H., Angew. Chem. Int. Ed. Engl. 33 (1994) 101.
8. Sprengard, U., Kretzschmar, G., Bartnik, E., Hüls, C. and Kunz, H., Angew. Chem. Int. Ed. Engl. 34 (1995) 990.

N-urethane protected carboxyanhydrides (UNCAs) in amino acid and peptide chemistry

J.-A. Fehrentz[a], C. Pothion[a], C. Devin[a], M. Paris[a], P. Chevallet[a], F. Winternitz[a], A. Loffet[b] and J. Martinez[a]

[a]Laboratoire des Aminoacides, Peptides et Proteines (LAPP), ESA 5075, CNRS-Universités Montpellier I & II, Faculté de Pharmacie, 15 av. C. Flahault, F-34060 Montpellier, France [b]Isochem, F-91710 Vert le Petit, France

Introduction

UNCAs are very reactive amino acid derivatives [1]. They have been used with success in SPPS and have demonstrated their usefulness in solution peptide synthesis [2]. Furthermore, we have shown that UNCAs can be considered as starting material for the synthesis of various amino acid derivatives.

Results and Discussion

Chemoselective reduction of UNCAs by sodium borohydride in the presence of water leads in quantitative yield to the corresponding β-amino alcohols [3]; no racemisation occurred during the reduction.

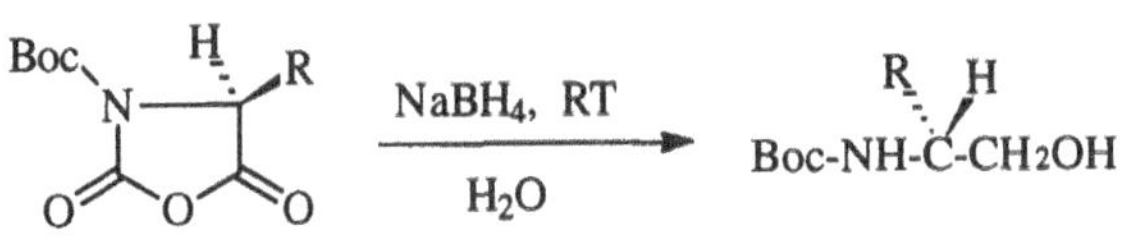

This reaction allowed us to obtain in a one-step synthesis the desired β-amino alcohols in good yields (80 to 96%). Reaction of UNCAs with Meldrum's acid in the presence of a tertiary amine followed by the intramolecular cyclization yields to enantiomerically pure tetrameric acid derivatives, which are precursors of γ-amino β-hydroxy-acids (statine

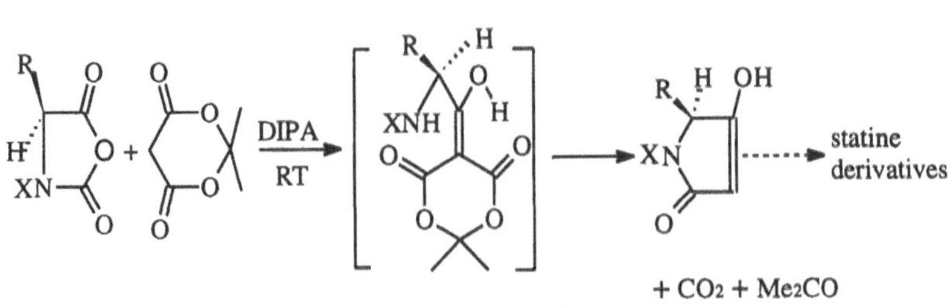

derivatives); this reaction is very simple, fast and proceeds from commercially available reagents. It was successfully performed with various Z, Boc or Fmoc UNCAs [4].

Reduction of UNCAs by bulky hydrides, i.e. lithium aluminium tris(tert-butyloxy)-hydride or lithium aluminium tris[(3-ethyl-3-pentyl)oxy]-hydride (LTEPA), yields the corresponding N-urethane protected α-amino aldehydes [5].

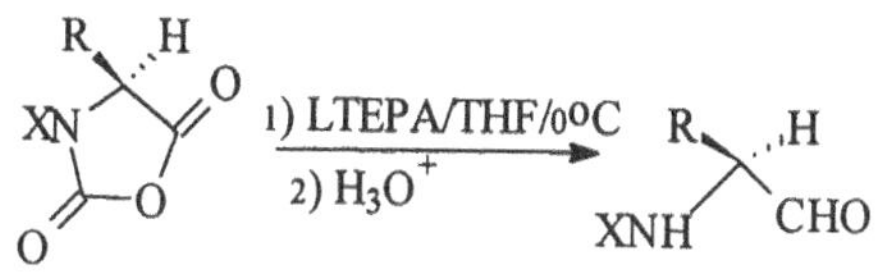

This simple reaction allows in a one-step procedure the synthesis of enantiomerically pure α-amino aldehydes from commercially available material in good yields (50 to 90%). UNCAs react smoothly and efficiently with phosphoranes to produce the corresponding keto phosphoranes in excellent yield [6]. These derivatives can lead to the vicinal tricarbonyl compounds by subsequent oxidation with ozone or with [bis(acetoxy)-iodo-]-benzene. These derivatives are potential inhibitors of serine proteases, or starting materials for the synthesis of various natural compounds.

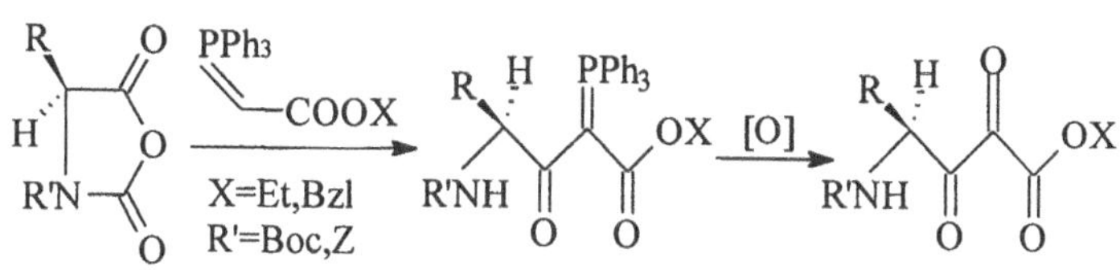

UNCAs react promptly with primary and secondary alcohols and thiols in the presence or absence of a base. In the case of tertiary alcohols, no reaction occurred when UNCAs were dissolved in the alcohol. However, when placed in tert-butanol as solvent, UNCAs (Boc or Z protected) in the presence of potassium bicarbonate and molecular sieves can smoothly lead to the corresponding tert-butyl esters. Reaction was performed at 45°C, for 45 hours. Yields are of about 70% and no significant racemization occurred during the reaction [7].

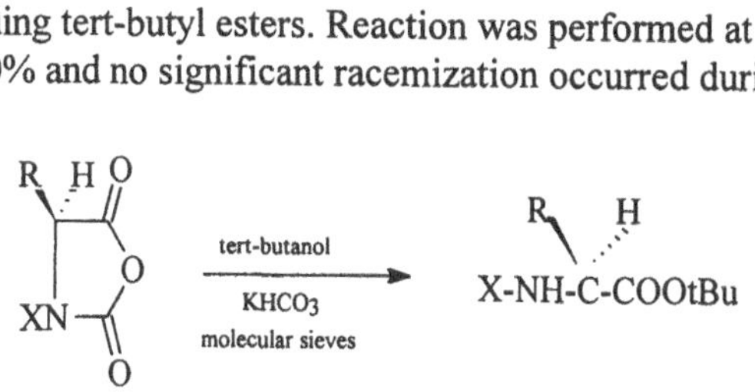

Dissolved in an anhydrous solvent in the presence of a tertiary amine i.e. DBU (1,8-diaza-bicyclo [5.4.0]undec-7-ene), in equimolar or catalytic quantities, UNCAs react promptly to give pyrrolidine-2,4-diones. The absolute configuration of the asymmetric carbons is under investigation by high field NMR studies in our laboratory. However, we found that

NMM can be used as base in the presence of UNCAs without any problem in solvents currently used in peptide synthesis [8].

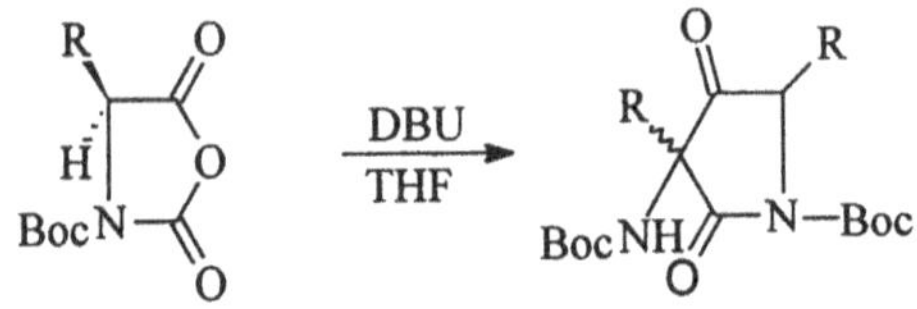

Conclusion

UNCAs are not only useful in peptide synthesis but also for the preparation of amino acid derivatives which are important intermediates in pseudo-peptide or peptoid chemistry. Reactions with UNCAs are characterized by their simplicity, efficiency and usually high yield. Various other reactions with UNCAs are under investigation in our laboratory.

References

1. Fuller, W.D., Cohen, M.P., Shabankareh, M. and Blair, R.K ., J. Am. Chem. Soc., 112 (1990) 7414. and Xue, C. and Naider, F., J. Org. Chem., 58 (1993) 350.
2. Rodriguez, M., Califano, J.C., Loffet, A . and Martinez, J., In Schneider, R.and Eberle, A. (Eds.) Peptides (EPS 1992), ESCOM Leiden, The Netherlands, 1993, p.233.
3. Fehrentz, J.A., Califano, J.C., Amblard, M., Loffet, A .and Martinez, J., Tetrahedron Letters, 35 (1994) 569.
4. Fehrentz, J.A., Bourdel, E., Califano, J.C., Chaloin, O., Devin, C.; Garrouste, P.; Lima-Leite, A.C., Llinares, M., Rieunier, F., Vizavonna, J. Winternitz, F. Loffet, A. and Martinez, J. Tetrahedron Letters, 35 (1994) 1557.
5. Fehrentz, J.A., Pothion, C., Califano, J.C., Loffet, A .and Martinez, J., Tetrahedron Letters, 35 (1994) 9031.
6. Fehrentz, J. A., Genu-Dellac, C., Amblard, M., Winternitz, F., Loffet, A. and Martinez, J., J. Peptide Science, 1 (1995) 124.
7. Chevallet, P., Fehrentz, J.A., Kiec-Kononowicz, K., Devin, C., Castel, J., Loffet, A. and Martinez, J., Lett. Pep. Sci., 2 (1995) 297.
8. Pothion, C., Fehrentz, J.A., Aumelas, A., Loffet, A. and Martinez, J., Tetrahedron Letters,37 (1996) 1027.

Immobilization of thermolysin for peptide synthesis with a high activity and stability

H. Eckstein, D. Hüttner and Xu Zheng
Institute of Organic Chemistry, University of Tübingen, Auf der Morgenstelle 18, D-72076 Tübingen, Germany

Introduction

Thermolysin (TLN) is an endoprotease and is used substantially for synthesising peptides (e.g. for aspartame precursors [1]) via the thermodynamic approach. For peptide synthesis, the use of immobilised enzymes is advantageous because of the possible continuous reaction in a flow-through reactor. Beyond this, the synthesised peptide will not be contaminated with the enzyme. After the coupling the immobilised enzyme can be removed by filtration and used again in later coupling steps [2].

In this investigation we describe the optimisation of the immobilization of TLN on three carriers, with and without some inhibitors and substrates during the immobilization.

Results and Discussion

The immobilization of TLN on Eupergit C (Röhm Pharma, Weiterstadt, Germany) is achieved by washing the polymer with water and then with 1 M phosphate buffer, pH 7.5. The wet polymer is added to a suspension of TLN in the same buffer. The suspension is rotated for 24 hours at room temperature. The optimal ratio is found to be 6 mg TLN for 150 mg Eupergit C. A higher concentration of TLN does not improve the activity. The immobilisate can be stored at -20°C until use.

The other carriers investigated are the aminosilica gels LiChroprep-amino (Merck, Darmstadt, Germany) and Polygosil-amino (Macherey-Nagel, Duren, Germany), which are activated with glutardialdehyde. The immobilization procedure is mainly the same as already reported [3].

The activity of the immobilised TLN is tested by the hydrolysis rate of Z-Asp-Phe-NH_2 at pH 8.5 and is determined by HPLC (sorbent: Nucleosil C4, 5µ (Macherey-Nagel, Duren, Germany); column: 100x2 mm; solvent: 40%B (A = 0.05 M NH_4OAc, pH 6.5, B = 80% MeOH/0.05 M NH_4OAc); flow rate: 0.3 ml/min).

In some cases higher active enzyme polymers are obtained by adding inhibitors or substrates. Both can protect the active site of the enzyme during the immobilization procedure to some extent. From a list of already published inhibitors [4] and substrates [5] we tried several compounds. From the tested inhibitors no improvement in the activity is observed. However, the activity is improved markedly by addition of Z-Ala during the immobilization procedure and by more than 20% in the case of PheOMe·HCl.

The biological stability of all immobilisates is very high as can be seen from Fig. 1. In this study the substrates Z-Phe, Z-Ala and Phe-OMe·HCl are used as additives.

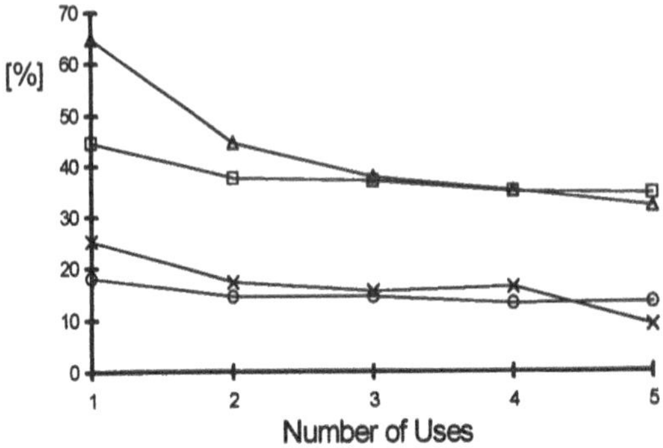

Fig. 1. Activities of TLN-Eupergit obtained by addition of Z-Phe (○), Z-Ala (□) and Phe-OMe•HCl (▵) during the immobilization of TLN and without substrate (×) being used up to five times.

The highest decrease of the activity is observed during the first use. Then the activity decreases only slowly or maintains nearly constant. In this respect the TLN-Eupergit obtained with the addition of Z-Ala (□) hydrolyses 35% of the test peptide after five uses. This polymer has then a higher activity in comparison to the TLN-Eupergit, obtained by addition of PheOMe·HCl, which had a much higher initial activity. Its activity is four times higher as without any admixture.

From all three carriers, the highest activity shows TLN-Polygosil, which hydrolyses in 60 min. 93% of the test peptide. The TLN-Eupergit hydrolyses 82%. The poorest activity has TLN-LiChroprep.

TLN-Polygosil is used for synthesis of the aspartame precursor Z-Asp-Phe-OMe in organic solvent. Ethyl acetate is saturated with 0.1M Tris·HCl (pH 7.5). 1 ml of a solution 160 mM of H_2N-Phe-OMe and 80 mM of Z-Asp is mixed with 300 mg of the wet polymer. Within 48 h 90% dipeptide is formed (HPLC control).

This demonstrates the efficiency of immobilised TLN in aqueous and organic solvents.

References

1. Isowa, Y., Ohmori, M., Ichikawa, T., Mori., K., Nonaka, Y., Kihara, K., Oyama, K., Satoh, H. and Nishimura, S., Tetrahedron Lett. (1979) 2611.
2. Eckstein, H. and Renner, H.J., In Maia, H.L.S. (Eds.) Peptides 1994 (Proceedings 23nd European Peptide Symposium), ESCOM, Leiden 1995, p. 230.
3. Cramer, S. M. and Horvath, C., Biotechnology and Bioengineering, 33 (1989) 344.
4. Zoller, H., In Handbook of Enzyme Inhibitors 2nd edition, VCH Weinheim 1993, p. 465.
5. Oka, T. and Morihara, K., J. Biochem. 88 (1980) 807.

Synthesis of polypeptide by a thioester method

Toru Kawakami, Shuji Kogure and Saburo Aimoto
Institute for Protein Research, Osaka University, 3-2 Yamadaoka, Suita, Osaka 565, Japan

Introduction

We have been developing a method of protein synthesis using peptide thioesters as building blocks and have reported syntheses of proteins by the method [1-5]. The synthetic procedure, however, becomes complicated, if a protein contains cysteine residues, since it becomes necessary to prepare a segment having benzyl-type protecting groups using an Npys solid-phase method and which are stable to silver ions [5]. The acetamidomethyl (Acm) group represents a more ideal blocking group for the mercapto group of a cysteine residue, since it is more easily prepared using a solid-phase method by Boc-strategy. If we could find the reaction conditions which allow a coupling reaction to proceed at a reasonable rate and under which the Acm group is stable toward silver ions, a simple and general method could be developed for the synthesis of cysteine-containing proteins.

This paper describes a method for the preparation of a cysteine-containing polypeptide using Cys(Acm)-containing peptide thioesters. To demonstrate the usefulness of the developed method, human adrenomedullin [6] consisting of 52 amino acid residues was synthesized. The amino acid sequence of human adrenomedullin is shown in Fig. 1, in which an arrow indicates the site of segment coupling. We also describe a method to convert a Cys(Acm)-containing peptide to the corresponding disulfide form of the peptide under basic conditions in the presence of silver ions and water.

Results and Discussion

An Acm group on a cysteine residue is stable under the HF treatment conditions. Thus, cysteine containing peptide thioesters can be easily prepared by a Boc solid-phase synthesis on an automatic peptide synthesizer.

The *N*-terminal peptide thioester segment **1**, Boc-[Cys(Acm)^{16}]-adrenomedullin(1-19)-SCH_2CH_2CO-β-Ala-NH_2, was prepared according to the scheme as shown in Fig. 2. The peptide was prepared by using Boc-Cys(Acm) for a cysteine residue according to the protocol of system software version 1.40 NMP/HOBt *t*-Boc (Applied Biosystems Inc., Foster City, CA). After the completion of chain elongation cycles, a protected peptide resin was treated with anhydrous HF containing 1,4-butanedithiol and anisole to give a crude product, which was purified by reverse phase high performance liquid chromatography (RPHPLC). The terminal amino group of the purified peptide was protected by a Boc group treatment with Boc-ONSu in the presence of DIEA to give partially protected peptide thioester **1** in 13% yield based on the Gly residue in the starting resin. Peptide **2**, [Cys(Acm)^{21}, Lys(Boc)^{25,36,38,46}]-adrenomedullin(20-52), was prepared from a peptide obtained by an Fmoc solid-phase method according to the protocol of FastMoc 0.25 ΩMonPrevPk (Applied Biosystems Inc.) in 17% yield based on the amino group in the starting resin. Peptide **2**, of course, can be easily prepared by a Boc solid-phase method as well.

Thioester **1** (1.7 μmol) and peptide **2** (1.7 μmol) were added to a solution containing $AgNO_3$ (5.1 μmol), HOOBt (52 μmol), and DIEA (34 μmol) in DMSO (350 μl) and the resulting solution was stirred for 24 h. Peptide **3**, Boc-[Cys(Acm)^{16,21}, Lys(Boc)^{25,36,38,46}]-adrenomedullin(1-52) (1.25 μmol, 73%) was isolated by RPHPLC and freeze-dried to give a powder. The elution profile of the reaction mixture is shown in Fig. 3. An aliquot of the

```
1                                      10
Tyr-Arg-Gln-Ser-Met-Asn-Asn-Phe-Gln-Gly-Leu-Arg-Ser-Phe-Gly-
                   20                                     30
Cys-Arg-Phe-Gly-Thr-Cys-Thr-Val-Gln-Lys-Leu-Ala-His-Gln-Ile-
               ↑                       40
Tyr-Gln-Phe-Thr-Asp-Lys-Asp-Lys-Asp-Asn-Val-Ala-Pro-Arg-Ser-
                   50     52
Lys-Ile-Ser-Pro-Gln-Gly-Tyr-NH2
```

Fig. 1. Amino acid sequence of adrenomedullin. An arrow indicates the site of segment coupling.

Boc-Gly-SCH_2CH_2CO-β-Ala-NH-resin

↓ ABI 430A Peptide synthesizer
System software version 1.40 NMP/HOBt *t*-Boc
End capping by acetic anhydride

Boc-Tyr(Br-Z)-Arg(Tos)-Gln-Ser(Bzl)-Met-Asn-Asn-Phe-Gln-Gly-Leu-Arg(Tos)-Ser(Bzl)-Phe-Gly-Cys(Acm)-Arg(Tos)-Phe-Gly-SCH_2CH_2CO-β-Ala-NH-resin

↓ 1) HF treatment
2) RPHPLC

Tyr-Arg-Gln-Ser-Met-Asn-Asn-Phe-Gln-Gly-Leu-Arg-Ser-Phe-Gly-Cys(Acm)-Arg-Phe-Gly-SCH_2CH_2CO-β-Ala-NH_2

↓ 1) Boc-OSu, DIEA
2) RPHPLC

Boc-Tyr-Arg-Gln-Ser-Met-Asn-Asn-Phe-Gln-Gly-Leu-Arg-Ser-Phe-Gly-Cys(Acm)-Arg-Phe-Gly-SCH_2CH_2CO-β-Ala-NH_2 (**1**)

*Fig. 2. A synthetic route of peptide **1**, Boc-[Cys(Acm)^{16}]-adrenomedullin(1-19)-SCH_2CH_2CO-β-Ala-NH_2.*

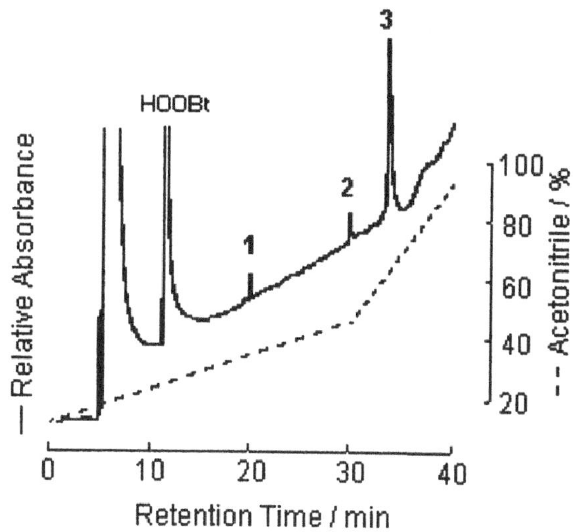

Fig. 3. RPHPLC elution profie of the reaction mixture of peptide 3.

purified peptide **3** (0.72 μmol) was treated with TFA containing 5% 1,4-butanedithiol for 1 h to give peptide **4**, [Cys(Acm)[16,21]]-adrenomedullin(1-52) (Fig. 4 (A)). The mixture was washed with ether, then lyophilized from aqueous acetonitrile. Peptide **4** obtained was treated with AgOTf in TFA [7] to remove Acm groups, but the Acm groups could not be removed. Based on the observation that Acm groups were removed during the segment coupling reaction using$AgNO_3$, HONp, and DIEA, we designed a route to remove Acm group under basic conditions in the presence of nucleophile. Thus, peptide **4** was dissolved into water (0.20 ml), to which was added a solution of $AgNO_3$ (3.6 μmol) and DIEA (11 μmol) in DMSO (70 μl). The resulting solution was stirred for 2 h to form peptide **5**, [Cys(Ag)[16,21]]-adrenomedullin(1-52) (Fig. 4 (B)). To the reaction mixture was added a mixture of 1 M HCl and DMSO (1 : 1, v/v, 3.0 ml), and the mixture was stirred for 24 h. The native form of adrenomedullin was isolated by RPHPLC in 57% yield based on peptide **3** (3.6 mg, 0.41 μmol) as shown in Fig. 4 (C). The mass number and amino acid analysis data agreed well with those of expected values.

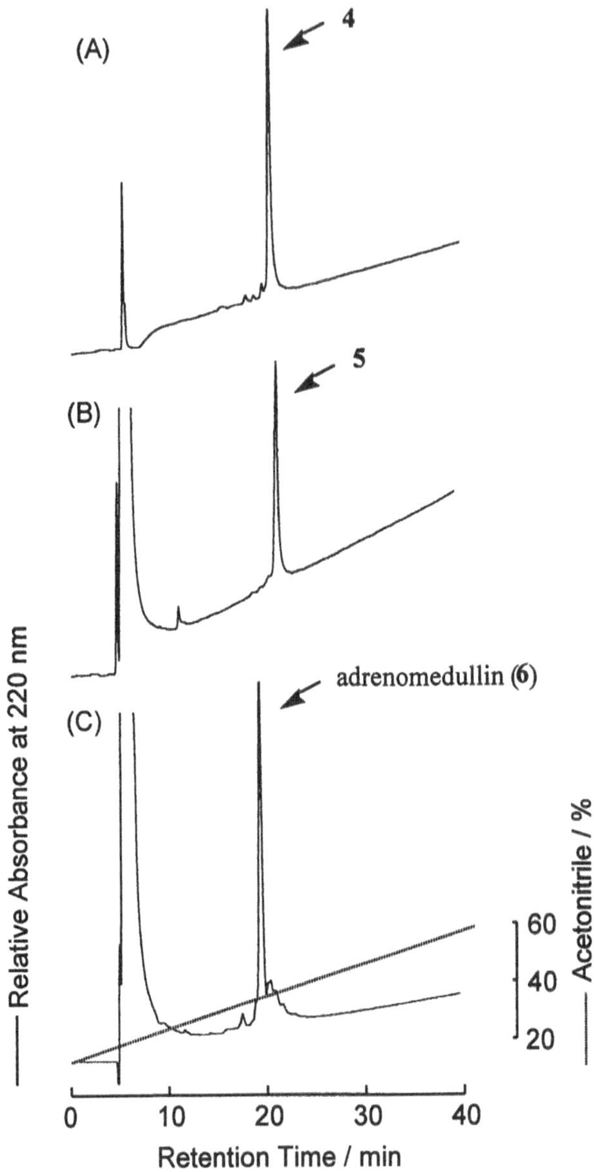

*Fig. 4. RPHPLC elution profiles of the reaction mixtures of the preparation of peptides **4** (panel A), **5** (panel B), and **6** (panel C), respectively. Each reaction mixture was analyzed by Cosmosil 5C_{18} (10x250 mm) by linear increase of acetonitrile concentration from 20 to 50% in 0.1% aq. trifluoroacetic acid over the period of 30 min at a flow rate of 2.5 ml/min.*

Acknowledgment

This research was partly supported by the grant-in-Aid for Scientific Research on Priority Areas No. 06276102 from the Ministry of Education, Science, Sports and Culture.

References

1. Hojo, H. and Aimoto, S., Bull.Chem. Soc. Jpn., 64 (1991) 111.
2. Hojo, H. and Aimoto, S., Bull.Chem. Soc. Jpn., 65 (1992) 3055.
3. Hojo, H, Kwon, Y.D., Kakuta, Y., Tsuda, S.,Tanaka, I., Hikichi, K. and Aimoto, S., Bull. Chem. Soc. Jpn., 66 (1993) 2700.
4. Hojo, H. and Aimoto, S., Bull. Chem. Soc. Jpn., 66 (1993) 3004.
5. Hojo, H., Yoshimura, S., Go, M.. and Aimoto, S., Bull. Chem. Soc. Jpn., 68 (1995) 330.
6. Kitamura, K., Kangawa, K., Kawamoto, M., Ichiki, Y., Nakamura, S., Matsuo, H. and Eto, T., *Biochem.* Biophys. Res. Commun., 192 (1993) 553 .
7. Fujiii, N., Otaka, A, Watanabe, T., Okamachi, A., Tamamura, H., Yajima, H., Inagaki, Y., Nomizu, M. and Asano, K., J. Chem. Soc., Chem. Commun., (1989)283.

Synthesis of amino acids selectively labelled with stable isotopes for application in peptide synthesis

U. Ragnarsson, L. Grehn, Y. Elemes and B. Fransson
Department of Biochemistry, University of Uppsala, Biomedical Center, P.O. Box 576, S-751 23 Uppsala, Sweden

Introduction

Amino acids labelled with stable isotopes such as ^{2}H, ^{13}C and ^{15}N were originally applied in metabolic studies and have subsequently played an important role also in many biosynthetic and mechanistic investigations. With the growing use of ^{13}C and ^{15}N NMR spectroscopy in structural studies of proteins, isotope labelling has often become a major tool to increase the sensitivity as the natural abundance of these nuclei is only 1.1 and 0.35%, respectively. As a result today more than 150 ^{13}C and/or ^{15}N isotopomers of the proteinogenic amino acids have been described [1], including both uniformly and selectively labelled species. The former can be obtained by hydrolysis of biomass and subsequent separation into pure components [2,3], whereas the latter are generally prepared by direct synthesis. Recently in our laboratory we have developed procedures for selective labelling of amino acids with stable isotopes for application in peptide synthesis and in this contribution they will be reviewed.

Results and discussion

Synthesis of ^{15}N-Labelled L-Amino Acids

Our procedure for the preparation of ^{15}N-labelled L-amino acids starts from commercial D-amino acids and leads directly to pure Boc-derivatives in high yield. The initial step is a classical transformation to hydroxy acid [4], followed by esterification of the carboxyl and conversion to triflate.

The ^{15}N nucleus as well as the N-protecting group are simultaneously introduced via the

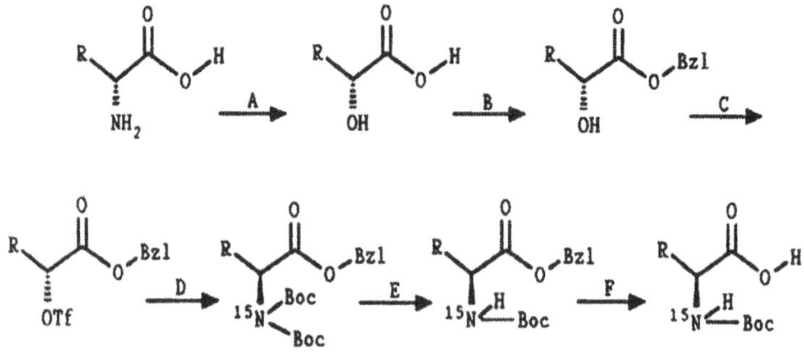

Scheme 1. Synthesis of Boc-L-[^{15}N]amino acids. A: HNO_2, B: BzlBr, C:$(CF_3SO_2)_2O$, D: $Li^{15}NBoc_2$, E: dil. TFA, F: hydrogenolysis.

lithium salt of bis-t-butyl (^{15}N)imidodicarbonate (5) prepared in situ. This reaction takes place with complete inversion of configuration as demonstrated by the sensitive Flec-method [6] and provides the product after two simple partial deprotection steps [7,8].

By variation of the imidodicarbonate, in principle other N-protected derivatives can be prepared similarly [7]. Instead of triflates, the Mitsunobu reaction can also be used [7], provided that the imidodicarbonate is acidic enough [9].

Synthesis of the Whole Set of Boc-[^{13}C,^{15}N]Glycines

A simple synthetic procedure for these isotopomers of glycine has been developed [10]. Reaction of the potassium salt of labelled and non-labelled bis-t-butyl imidodicarbonate with the three alternative ethyl [^{13}C]bromoacetates gave the corresponding fully protected species, which after consecutive partial deprotections furnished the Boc-derivatives in essentially quantitative yield. These isotopomers of glycine are required for asymmetric synthesis of other proteinogenic amino acids [11].

$$\mathrm{Boc_2{}^xNK + Br\text{-}{}^yCH_2\text{-}{}^zCO\text{-}OEt \longrightarrow Boc_2{}^xN\text{-}{}^yCH_2\text{-}{}^zCO\text{-}OEt}$$

$$\xrightarrow{A} \mathrm{Boc\text{-}{}^xNH\text{-}{}^yCH_2\text{-}{}^zCO\text{-}OEt} \xrightarrow{B} \mathrm{Boc\text{-}{}^xNH\text{-}{}^yCH_2\text{-}{}^zCOOH}$$

Scheme 2. Synthesis of Boc-[^{13}C,^{15}N]glycines. A:dil. TFA, B: NaOH. Compounds made (x/y/z):(15/13/13), (14/13/13), (15/12/13), (14/12/13), (15/13/12), (14/13/12).

Asymmetric Synthesis of [^{13}C,^{15}N]Labelled Amino Acids

Nowadays several excellent methods are available for asymmetric synthesis of α-amino acids [11]. In our work we have preferentially explored Oppolzer's method based on the use of (2R)-bornane-10,2-sultam as a chiral auxiliary [12] and used it according to Scheme 3 to prepare a number of [1,2-$^{13}C_2$,^{15}N]amino acids [13]. The final products had an optical purity generally exceeding 99% e.e. as determined by two independent chromatographic methods [6,14] in full agreement with those reported by Oppolzer et al. [12] and higher in our hand than that obtained by Schollkopf's bislactim method [15].

More recently, we have used Oppolzer's procedure to prepare also a few L-[2-^{2}H]- and, especially, [2-^{2}H,1,2-$^{13}C_2$,^{15}N]amino acids [16,17].

Conclusion

Many Boc-[^{15}N]amino acids can now be prepared from the corresponding non-labelled antipodes in high isotopic yield and very high optical purity with the help of [^{15}N]imidodicarbonates. [^{13}C]-Backbone-labelled amino acids with additional 2-^{2}H and ^{15}N nuclei are conveniently obtained by asymmetric synthesis using Oppolzer's

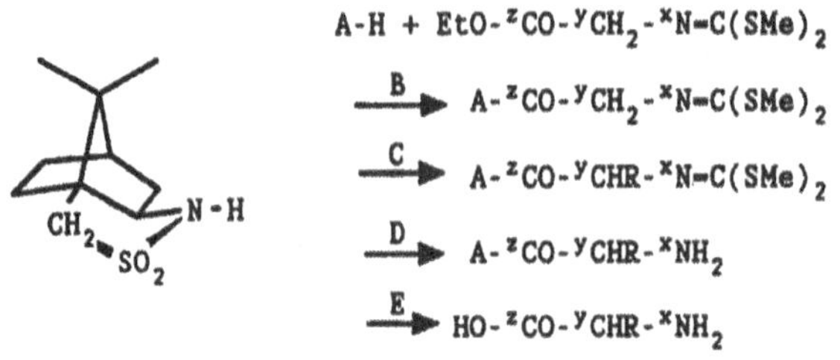

Scheme 3. Synthesis of backbone-labelled Boc-L-amino acids using Oppolzer's procedure A-H: Chiral auxiliary, B: Me_3Al, C: BuLi/RI, D: HCl, E:LiOH.

chiral auxiliary in excellent optical purity. Such amino acids have so far been used for a synthesis of one fully [1,2-$^{13}C_2$,^{15}N]-labelled peptide [18] and further such peptides are in progress.

Acknowledgments

This work was supported by the Swedish Natural Science Research Council and the National Board for Industrial and Technical Development. The costs for isotopes were defrayed by a grant from the Carl Trygger Foundation. Y.E. took part in the project on a postdoctoral scholarship from the Human Capital and Mobility Programme (EU, Brussels).

References

1. Ragnarsson, U., J. Peptide Sci., 1 (1995) 149.
2. Sogn, J.A., Craig, L.C. and Gibbons, W.A., Int. J. Pept. Protein Res., 6 (1974) 353.
3. LeMaster, D.M. and Richards, F.M., Anal. Biochem., 122 (1982) 238.
4. Brewster, P., Hiron, F., Hughes, E.D., Ingold, C.K. and Rao, P.A.D.S., Nature (London), 166 (1950) 179.
5. Degerbeck, F., Grehn, L. and Ragnarsson, U., Acta Chem. Scand., 47 (1993) 896.
6. Einarsson, S., Josefsson, B., Moller, P. and Sanchez, D., Anal. Chem., 59 (1987) 1191.
7. Degerbeck, F., Fransson, B., Grehn, L. and Ragnarsson, U., J. Chem. Soc., Perkin Trans. 1, (1992) 245.
8. Degerbeck, F., Fransson, B., Grehn, L. and Ragnarsson, U., J. Chem. Soc., Perkin Trans. 1, (1993) 11.
9. Koppel, I., Koppel, J., Degerbeck, F., Grehn, L. and Ragnarsson, U., J. Org. Chem., 56 (1991) 7172.
10. Grehn, L., Pehk, T. and Ragnarsson, U., Acta Chem. Scand., 47 (1993) 1107.
11. Williams, R.M. Synthesis of Optically Active α-Amino Acids, Pergamon Press, Oxford, 1989.
12. Oppolzer, W., Moretti, R. and Thomi, S. Tetrahedron Lett., 30 (1989) 6009.
13. Lankiewicz, L., Nyasse, B., Fransson, B., Grehn, L. and Ragnarsson, U., J. Chem. Soc., Perkin Trans. 1, (1994)2503.
14. Frank, H., Nicholson, G.J. and Bayer, E. J. Chromatogr. Sci., 15 (1977) 174.
15. Schollkopf, U., Pure Appl. Chem., 55 (1983) 1799.
16. Elemes, Y. and Ragnarsson, U., J. Chem. Soc., Perkin Trans. 1 (1996) 537.
17. Elemes, Y. and Ragnarsson, U., Chem. Commun., (1996) 935.
18. Nyasse, B., Grehn, L. and Ragnarsson, U., J. Chem. Soc., Chem. Commun., (1994) 2005.

Design and synthesis of fluorinated Phe-Gly dipeptidomimetics

W. Berts, R.C. Vollinga, K. Luthman and U. Hacksell
Department of Organic Pharmaceutical Chemistry, Uppsala Biomedical Centre, Uppsala University, Box 574, S-751 23 Uppsala, Sweden

Introduction

Isosteric replacement of amide bonds in biologically active peptides may prevent their proteolytic degradation and is a frequently used strategy in studies of structural mimicry [1]. In the design of amide bond isosteres several important factors have to be considered. The isosteres have to mimic the amide geometrically and electrostatically, and should also show similar conformational flexibility and hydrogen bonding properties. A previously reported Phe-Gly fluorovinyl isostere exhibits good mimicking ability [2]. Therefore, we have synthesized a series of novel unsaturated and saturated mono- and difluorinated Phe-Gly dipeptidomimetics to be used as building blocks in the synthesis of pseudopeptides.

Results and Discussion

The fluorinated derivatives were synthesized in several steps starting from the *trans*-Phe-Gly vinyl isostere **1**. The reaction sequence started with a stereo-selective epoxidation using *m*-CPBA to afford a diastereomeric mixture of epoxides [3,4]. A stereospecific ring opening using fluoride ion gave the allylic alcohols in good yield [3]. When alcohol **3** was reacted with DAST we obtained a mixture of fluorinated derivatives and the *cis*-aziridine **4** [5]. Aziridine **4** was then reacted with DAST to specifically afford the allylic fluoro derivative **5**. Antimony trichloride was used as a catalyst in the reaction [6]. The aziridine could be obtained from the pure alcohol isomer also by using Mitsunobu conditions. This reaction was cleaner and higher yielding than the DAST reaction. The *trans*-aziridine was obtained from the corresponding alcohol in 68% yield. This aziridine was then reacted with DAST to afford the epimer of **5** in 60% yield. The *cis*- and *trans*-aziridines show different reactivities in the reaction with DAST, the *trans* isomer being more reactive. The saturated fluoro derivative **6** was obtained in quantitative yield by hydrogenation of **5**.

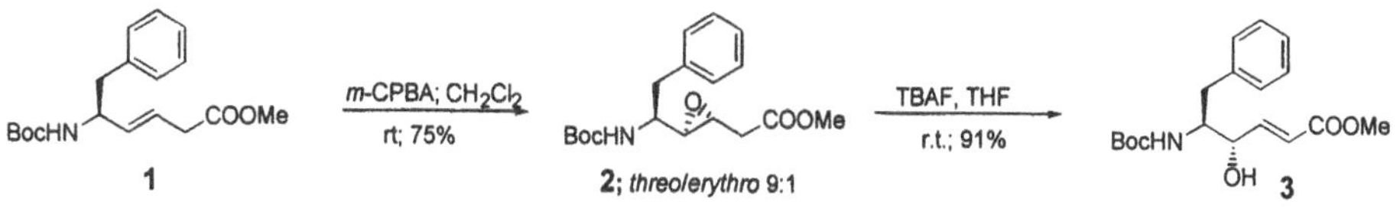

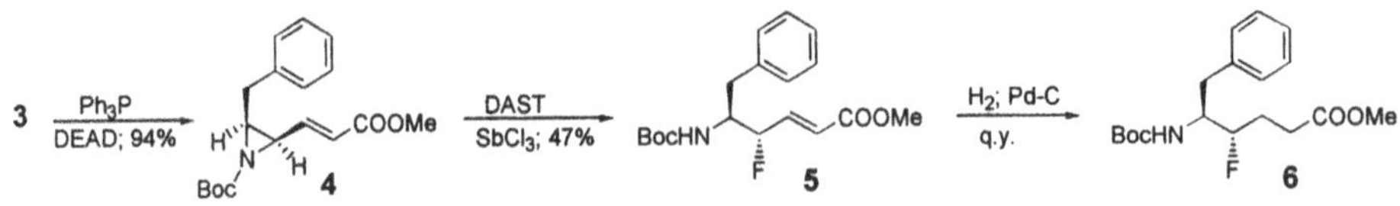

Geminal difluorinated compounds can be obtained from the corresponding ketones [7]. We therefore synthesized the unsaturated ketone **7** from an isomeric mixture of alcohols. However, the ketone did not react with DAST even after prolonged heating. We believe that the low reactivity is due to the stability of the conjugated system which decreases the Lewis basicity of the carbonyl oxygen. Selective hydrogenation of the double bond using Pd-C in pyridine afforded **8**, which could be converted to the difluoro derivative **9** by treatment with DAST.

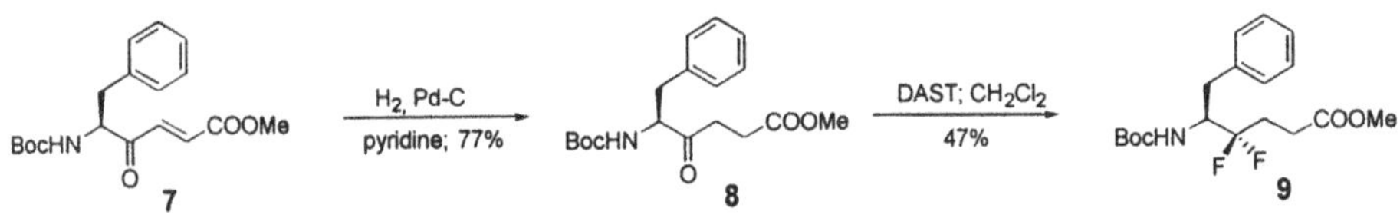

The mimicking ability of the synthesized derivatives will be evaluated after replacement of Phe-Gly dipeptidic moieties in biologically active peptides, e.g. substance P and dermorphin. Both these peptides contain Phe-Gly fragments in domains considered important for receptor binding and activation.

Acknowledgment

This work was supported by grants from The Swedish Board for Industrial and Technical Development (NUTEK) and from The Swedish Natural Science Research Council (NFR).

References

1. Spatola, A., In Weinstein, B. (Ed.) Chemistry and Biochemistry of Amino Acids, Peptides and Proteins, Vol. 7, Marcel Dekker Inc., New York, 1983, p 267.
2. Allmendinger, T., Tetrahedron Lett., 31 (1990) 7297 and 31 (1990) 7301.
3. Jenmalm, A., J. Org. Chem., 59 (1994) 1139.
4. Jenmalm, A.., J. Org. Chem., 60 (1995) 1026.
5. Li, L.-Y., Tetrahedron Lett., 33 (1992) 4487.
6. Robins, M. J. and Wnuk, S. F., Tetrahedron Lett., 29 (1988) 5729.
7. Bird, T.G.G.., J .Chem. Soc. Chem. Commun., (1979) 65.

New methods for synthesis of bis(cystine) peptide dimers

Lin Chen and George Barany
Department of Chemistry, University of Minnesota, 207 Pleasant Street, S.E., Minneapolis, MN 55455, U.S.A.

Introduction

Although peptide dimers frequently arise as unwanted by-products from syntheses directed at monomeric cyclic disulfide peptides, it is quite challenging to synthesize peptide dimers intentionally [1,2]. For the present work, we have demonstrated a new approach to generate homo and hetero peptide dimers (both parallel and antiparallel), using deamino-oxytocin and oxytocin systems as models.

Results and Discussion

The linear sequences were assembled by Fmoc solid-phase synthesis techniques, with orthogonal protection for the two β-thiols by appropriate combinations of S-[(N'-methyl-N-phenylcarbamoyl)sulfenyl] (Snm) [3], S-acetamidomethyl (Acm) [1], and S-2,4,6-trimethoxybenzyl (Tmob) [4] groups (Fig. 1). The first disulfide bonds were formed in solution by directed methods, and then the second disulfide bonds were formed, without purification of the intermediates, by iodine oxidation (Fig. 2). The methods were readily adapted to prepare a series of dimers in a convergent manner (common peptide-resin intermediates applied to prepare multiple products). Final yields of pure dimers (> 98% by HPLC) were in the 20 to 40% range (yield higher for homo than for heterodimers).

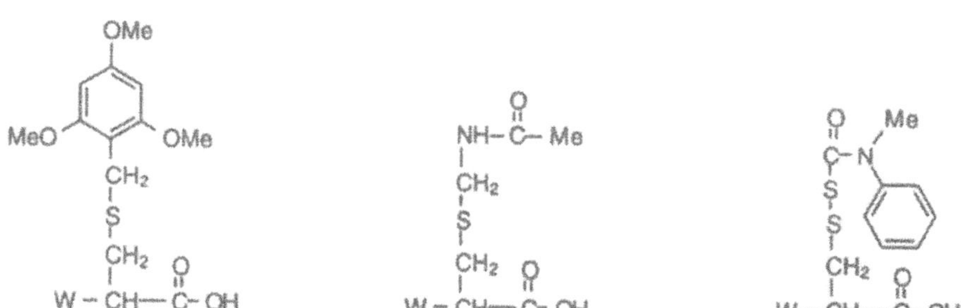

1 2 3

1a, W=H, Mpa(tmob)-OH
1b, W=Fmoc-NH,Fmoc-Cys(Tmob)-OH
2a, W=H, Mpa(Acm)-OH
2b, W=Fmoc-NH, Fmoc-Cys(Acm)-OH

3a, W=H, Mpa(snm)-OH
3b, W=Boc-NH, Boc-Cys(Snm)-OH
3c, W=Fmoc-NH, Fmoc-Cys(Snm)-OH

Fig. 1. Protected derivatives of cysteine and β-mercaptopropionic acid.

Fig. 2. Syntheses of parallel and antiparallel dimers of deamino-oxytocin.

Acknowledgments

We thank Dr. Jirina Slaninova, our collaborator on the biological aspects of this work, which will be reported elsewhere, and Dr. Edmund A. Larka and Mr. Sean Murray for outstanding contributions with mass spectrometric analyses. This work was supported by NIH grants GM 42722 and 43552.

References

1. Andreu, D., Albericio, F., Solé, N.A., Munson, M.C., Ferrer, M. and Barany, G., In Pennington, M.W. and Dunn, B.M. (Eds.) Methods in Molecular Biology, Vol. 35, Peptide Synthesis Protocols, Humana, Totowa, N.J., 1994, p. 91.
2. Munson, M.C., Lebl, M., Slaninov, J., and Barany, G., Peptide Res., 6 (1993) 155.
3. Schroll, A.L., and Barany, G., J. Org. Chem. 54 (1989) 244.
4. Munson, M.C., Garcia-Echeverria, C., Albericio, F, and Barany, G., J. Org. Chem. 57 (1992) 3013.

Ligation of laminin fragments onto a PEG dendrimer

Lei Huang[a], De-Xin Wang[a] and Shi-Jun Li[b]
[a] *Institute of Materia Medica, Chinese Academy of Medical Sciences, Beijing 100050, China*
[b] *Hospital of Peking Union Medical College, Chinese Academy of Medical Sciences, Beijing 100730, China*

Introduction

The graft of poly-ethylene glycol (PEG) onto proteins, peptide-haptens and other molecules usually produces some useful properties, such as enhanced stability, improved plasma half-life, increased solubility and resistance to proteolysis [1]. However low density of pharmacophore moiety in PEG grafted molecule probably is not as effective when compared to MAP molecule [2]. In this paper we describe the synthesis of a new macromolecule with MW up to 8.0 KDa, consisting of PEG and MAP structure by a site-specific reaction.

Results and Discussion

In order to ligate various compounds, it is necessary to use chemistry which is very mild and specific. Several kinds of structures, such as amide, oxime, hydrazone, Schiff base, thiazolidine, etc. can be used as the linkage. Thiazolidine structure [3] was chosen by us for its high site-specificity. PEG carrier, a moiety of the target compound 6, was functionalized by three step reaction (Fig.1). The other moiety of 6, an unprotected nonapeptide fragment of laminin, was prepared by SPPS on MBHA resin.

Compound 3 was identified by ninhydrin test (-) and 2,4-dinitro-phenylhydrazine (+). The formation of thiazolidine ring between 1,2-aminothiol of the N-terminal Cys of 5 and aldehyde group of 3 was accomplished in 0.02 M NaOAc buffer containing 0.008 M EDTA, pH 5 at 50 °C for 4 h [3]. A major advantage of this reaction was the use of nucleophile (1,2-aminothiol) which was the only reactive group towards the aldehyde group of 3. The other side-chain groups were excluded from the reaction by protonation at acidic pH. The final product 6 was purified by dialysis (cut off<1.2KDa) and confirmed by amino acid analysis. The structure integration of MAP and PEG into one molecule described in this paper is an attempt to develop a new compound, which not only has higher molecular weight than MAP and higher density of peptide-hapten than conventional PEG modifier but may also possess a new conformational feature.

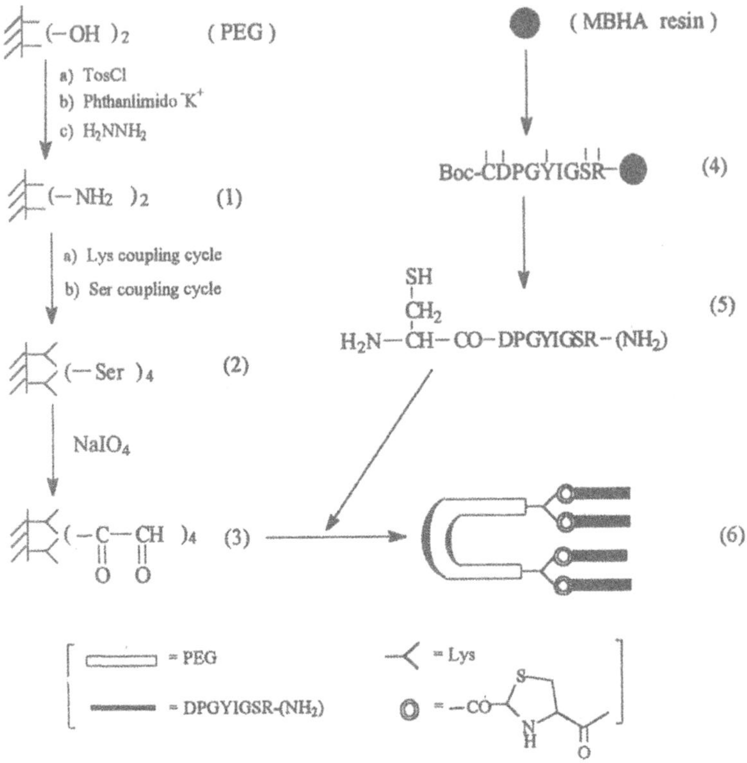

Fig. 1. Synthetic route of PEO dendrimeric peptide.

References

1. Nucci, M. L., Shorr, R. and Abuchowski, A., Adv. Drug Del. Res., 6 (1991) 133.
2. Tam, J. P., Proc. Natl. Acad. Sci. USA, 85 (1988) 5409.
3. Spetzler, J. C. and Tam, J. P., Int. J. Pept. Prot. Res., 45 (1995) 78.

Design and total synthesis of two peptide analogs of actinomycin D: 5,5'-Val_2-AMD and 2,2'-Phe_2-AMD

Rui Wang[a], Jing-Man Ni[b], Xin-Fu Pan[b], Xiao-Wu Yang[a] and Xiao-Yu Hu[a]

[a]Department of Biology and [b]Department of Chemistry, State Key Laboratory of Applied Organic Chemistry, Lanzhou University, Lanzhou 730000, China

Introduction

Actinomycin D (AMD, Fig.1) is used clinically for treating a limited number of highly malignant tumors such as Wilms' tumor, gestational choriocarcinoma. Although it possesses valuable antitumor activities, its high cytotoxicity and inactivity towards some tumors have limited its wide use and prompted the search for modified AMD. On the basis of the results of AMD-DNA binding model [1], we designed and synthesized 5,5'-Val_2-AMD and 2,2'-Phe_2-AMD.

Results and Discussion

The title compounds were synthesized from **3** in 13 steps with overall yields of 27%, 17%, respectively. The structures were identified with ^{1}HNMR, 2DNMR, MS and HRMS.

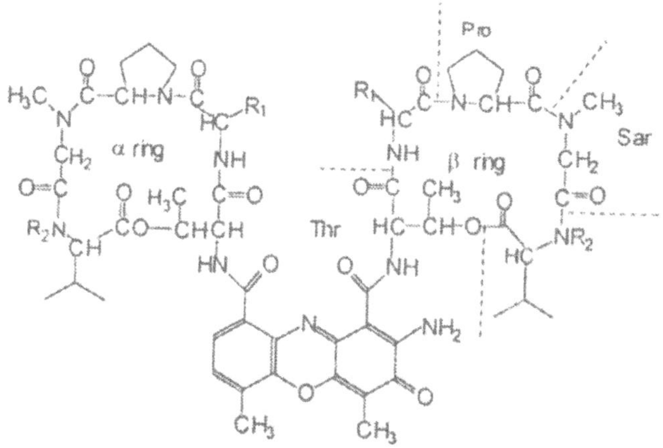

	$R_1 = CH(CH_3)_2$	$R_2 = CH_3$	AMD
1.	$R_1 = CH(CH_3)_2$	$R_2 = H$	5,5'-Val_2-AMD
2.	$R_1 = PhCH_2$	$R_2 = CH_3$	2,2'-Phe_2-AMD

Fig. 1. Structures of AMD and its analogs.

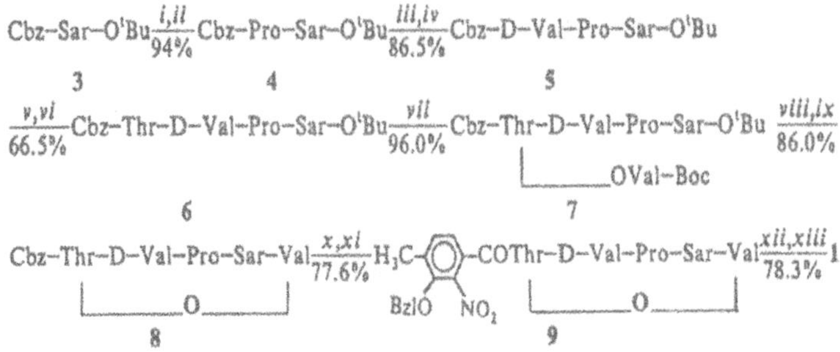

Reagents: i)H_2,10%Pd / C ii)Cbz–Pro,DCC iii)H_2,10%Pd / C iv)Cbz–D–Val,DCC v)H_2, 10%Pd / C vi)Cbz–Thr,DCC vii)Boc–Val,DCC,DMAP viii)TFA,0℃ ix)TEA,BOP–Cl x)H_2, 10%Pd / C xi)BMNBA,DCC xii)H_2,Pd / C xiii)$K_3Fe(CN)_6$

Scheme 1. Total synthesis of 5,5'-Val$_2$-AMD.

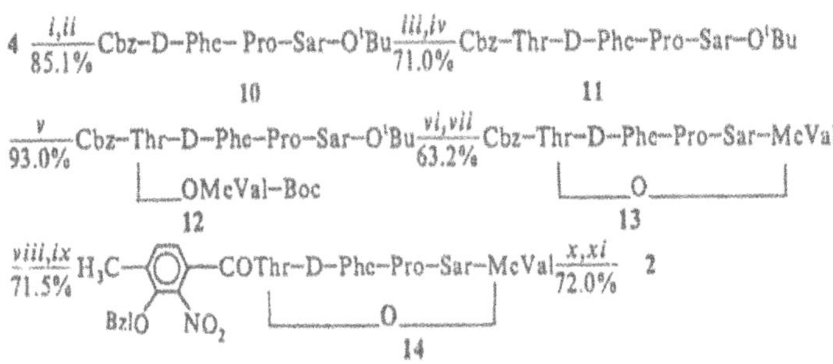

Reagents: i)H_2,10%Pd / C ii) Cbz–D–Phe, DCC iii)H_2, 10%Pd / C iv) Cbz–Thr, DCC v)Boc–MeVal, DMAP, DCC vi)TFA, 0℃ vii)TEA, Bop–Cl viii) H_2, 10% Pd / C ix) BMNBA, DCC x)H_2,10%Pd / C xi) $K_3Fe(CN)_6$

Scheme 2. Total synthesis of 2,2'-Phe$_2$-AMD.

Acknowledgments

This work was supported by grants from the National Natural Science Foundation of China (No. 29442010, No. 29502007), Fok Ying Tung Education Foundation and the State Education Commission of China.

References

1. Kamitori, S. and Takusagawa, F., J. Mol. Biol., 225 (1992) 445.

2,2-Difluoroboroxazolidin-5-ones: A novel approach to selective side-chain protections of serine and threonine by *tert*-butyl or benzyl groups

Ji-Dong Wang[a], Yoshio Okada[b] and Wei Li[a]
[a]*Department of Molecular Biology, Jilin University, Changchun 130023, China*
[b] *Faculty of Pharmaceutical Sciences, Kobe Gakuin University, Kobe 651-21, Japan*

Introduction

In peptide synthesis, Boc and Fmoc chemistries are routine methods to obtain desired peptides. Traditionally, the side chain hydroxyl functions in Ser and Thr have been protected as benzyl (Bzl) ethers based on Nα-Boc protection, and as *tert*-butyl (*t*-Bu) ethers in combination with Fmoc group as the Nα-protecting group. More recently, Fmoc-Ser(Bzl) and Fmoc-Thr(Bzl) have been employed in convergent solid phase peptide synthesis. However, methods for the preparation of these side-chain protected compounds are cumbersome [1- 4]. We have obtained **1** (Scheme) from aspartic acid and glutamic acid, and reported the synthesis of Asp(*t*-Bu) and Glu(*t*-Bu) using **1** as intermediates which give a simultaneous protection of both α-function groups and are stable to Lewis acid catalysts [7]. In this paper, we describe an efficient and practical one-pot synthesis of the Bzl and *t*-Bu-based side-chain protected derivatives of serine and threonine.

Results and Discussion

Formation, Deprotection and Properties of 1

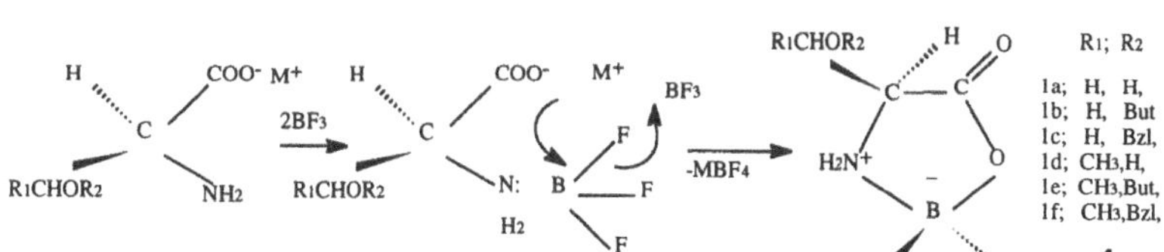

1

Mono alkali metal salt of amino acid [7] was treated with more than 2 equivalents of $BF_3.Et_2O$ in tetrahydrofuran (THF) to give **1**. The reaction mechanism is described in the Scheme. In aqueous solution, the deprotection is accelerated in the presence of bases. The chemical shifts of the two different fluorine atoms in **1a** (from trifluoroacetic acid) were -74.92 and -75.41 p.p.m. Compounds **1a** and **1d** were used in the preparation of side-chain protected Ser, Thr derivatives directly without further purification.

*Preparation of Ser(*t*-Bu) and Thr(*t*-Bu)*

Compound **1a** or **1d** was reacted with isobutylene in dioxane under the catalysis of

BF_3/H_3PO_4 to give **1b** and **1e** respectively. After hydrolysis in basic aqueous solution, Ser(*t*Bu) and Thr(*t*Bu) were obtained. Nonionic resin Amberlite XAD-2 or 4 are useful in the separation of amino acid derivatives from unsubstituted amino acids and inorganic salts in neutral aqueous solutions.

Preparation of Ser(Bzl) and Thr(Bzl)

Compound **1a** or **1d** was reacted with benzyl trichloroacetimidate, with BF_3 as the catalyst, in dioxane to give **1c** or **1f**, respectively. The desired compounds can be obtained after deprotection of the intermediates.

Conclusion

In sum, the present study demonstrates that **1** are highly effective, convenient, and in particular, inexpensive intermediates for simultaneous protections of both α-amino and α-carboxy groups in α-amino acids. The new method has simplified the tedious procedures of side-chain protection of Ser and Thr, and will pave the way for further application of this technique to many variants of side-chain modified amino acids, especially O-glycosylated amino acids [8].

References

1. Wunsch, E. and Jentsch, J., Chem. Ber, 97 (1964) 2490.
2. Murase, Y., Okawa, K. and Akabori, S., Bull. Chem. Soc. Jpn., 33 (1960) 123.
3. Sugano, H. and Miyoshi, M., J. Org. Chem., 41 (1976) 2352.
4. Mizoguchi, T., Levin, G., Woolley, D.W. and Stewart, J. M., J. Org. Chem., 33 (1968) 903.
5. Golubev, A., Sewald, N. and Burger, K., Tetrahedron Lett., 36 (1995) 2037.
6. Nefkens, G. H. L. and Zwanenburg, B., Tetrahedron., 39 (1983) 2995.
7. Wang, J. D., Okada, Y., Wang, Z. M., Wang, Y. H. and Li, W., Tetrahedron Lett., 1966.
8. Richard, R. S. and Josef, M., Angew. Chem. Int. Ed. Engl., 91 (1958) 1049.
9. Chang, C. D, Waki, M., Ahmad, M., Meienhofer, J., Lundell, E. and Haug, J, D., Int. J. Pept. Protein Res., 15 (1980) 59.
10. Hayakawa, T., Harada, K. and Fox, S, W., Bull. Chem. Soc. Jpn., 39 (1966) 391.
11. Fields, G. B. and Noble, R. L., Int. J. Peptide Protein Res., 35 (1990) 161.

Signalling approaches to inhibition of cellular proliferation

A.M. Doherty, W.L. Cody, D.M. Leonard, M. Creswell, G.L. Bolton, D. McNamara, S.R. Eaton, K.R. Shuler, J. Scholten, J.S. Sebolt-Leopold and R.L. Panek
Parke-Davis Pharmaceutical Research, Division of Warner-Lambert Company, Department of Medicinal Chemistry, Ann Arbor, MI 48105, U.S.A.

Introduction

There has been considerable progress in elucidating the mechanisms by which extracellular signals are transduced via cell surface receptors to trigger changes in gene expression determining the growth and differentiated state of the cell. Efforts to understand and target the mechanisms that underlie the development of abnormal proliferative diseases including vascular injury and cancer are under intense study [1,2]. We have targeted several strategies to inhibit cellular proliferation including blockade of platelet derived growth factor (PDGF) mediated mitogenic signalling and secondly with a downstream target by interfering with ras protein function via the enzyme ras farnesyl transferase (FTase) [1,2]. Strategies to develop potent cellular acting peptide inhibitors of the association of the C-terminal SH2 domain of the p85 subunit of phosphatidyl 3-kinase (PI_3 kinase) with the PDGF-βreceptor from the phosphorylated pentapeptide $Tyr(PO_3H_2)^{751}$-Val-Pro-Met-Leu (IC_{50} = 0.67 pM) have led to potent tetra- and tripeptide inhibitors. In the second signalling strategy, development of potent peptidomimetic inhibitors of the enzyme FTase are presented. Truncated tri- and dipeptides and peptidomimetics with cellular activity derived from the pentapeptide lead PD 083176 (Cbz-His-Tyr(OBn)-Ser(OBn)-TrpDAla-NH_2) (IC_{50}=17 nM) have been discovered.

Results and Discussion

From the pentapeptide $Tyr(PO_3H_2)^{751}$-Val-Pro-Met-Leu, which inhibits the association of PDGF β-receptor wth p85 C-terminal SH2 domain of PI_3 kinase with an IC_{50} of 0.67μM, C-terminal deletion of Leu^{755} resulted in a ten-fold loss of activity. However, N-terminal acetylation and C-terminal amidation enhanced the activity of the phosphorylated tetrapeptide Ac-Tyr(PO_3H_2)-Val-Pro-Met-NH_2 (IC_{50} = 0.24μM) [3,4]. Pseudoisosteric substitution of norleucine (NLe) for methionine led to an only two-fold loss of affinity compared with the parent molecule. When the NLe sidechain was transposed to the nitrogen of the amide to potentially increase stability and Pro was replaced with Ala for synthetic ease, this gave rise to a compound wth IC_{50} of 2.5μM. Subsequent replacement of the C-terminus with a long lipophilic group led to the discovery of the peptidomimetics Ac-Tyr(PO_3H_2)-Val-Ala-N$(C_5H_{11})_2$ and Ac-Tyr(PO_3H_2)-Val-Ala-N$(C_6H_{13})_2$ with improved activities (IC_{50} of 0.22 and 0.077μM) in inhibiting the PDGFβR: PI_3 kinase interaction [5]. In order to evaluate the potential therapeutic utility of these phosphorylated peptides, the difluoro-phosphonate derivatives (CF_2Pmp) of the penta- and tetrapeptide were synthesized and found to maintain high affinity while imparting cellular stability. These analogues [CF2Pmp^{751}Val-Pro-Met-Leu, IC_{50}= 1.2μM and CF_2Pmp^{751}-Val-Pro-Met-NH_2, IC_{50} = 1.7μM] exhibited micromolar cellular activity in inhibiting the PDGFβR: PI_3 kinase complex formation in rat smooth muscle cells (IC_{50}s of 40 and 38μM respectively).

In the strategy to discover inhibitors of ras farnesyl transferase, truncation of the pentapeptide (Cbz-His-Tyr(OBn)-Ser(OBn)-Trp-DAla-NH_2), derived from compound library screening led to the tripeptide, CBZ-D-His-Tyr(OBn)-Ser(OBn)-CO_2Me as a potent, selective inhibitor of ras FTase (IC_{50} = 0.37μM) [6]. Incorporation of D-His and N-methyl Tyr(OBn) led to the modified dipeptide PD 154309 (IC_{50} = 0.4μM) [7]. Transposition of the Tyr side chain of the tripeptide from the α-carbon to the adjacent nitrogen afforded PD 152440 which also showed good inhibitory activity (IC_{50} = 0.4μM) [8]. Both PD 154309 and PD 152440 inhibit ras processing in H-ras transformed NIH/3T3 cells with a minimum effective dose of 1μM. In the PD 154309 series, substitutions that lead to good activity both against the enzyme and in cells included among others the L-His analog (IC_{50}= 0.70μM), incorporation of $NHCH_2CH_2Ph$(2-Cl) at the C-terminus (IC_{50}= 0.93μM) and incorporation of D-NMe-His (IC_{50}= 1.2μM). A number of C-terminal analogs of PD 152440 were prepared; a lipophilic side chain was required to maintain *in vitro* activity. The location of the oxygen in the side chain was not critical for activity, however, chain length was important. An examination of substitution on the N-benzyl side chain revealed more stringent requirements for maintaining activity. However, the biphenyl derivative with Ph replacing OBn led to a compound with good *in vitro* and cellular potency (IC_{50} FTase = 0.36μM).

References

1. Seedorf, K. And Metabolism, 44 (1995) 24.
2. Khosravi-Far Roya and Der, C.J., Cancer and Metastasis Reviews, 13 (1994) 67.
3. Ramalingam, K., Eaton, S.R., Cody, W.L., Lu, G.H., Panek, R.L., Waite, L.A., Decker, S.J., Keiser, J.A., and Doherty, A.M., Bioorg. Med. Chem., 3 (1995) 1263.
4. Eaton, S., Ramalingam, K., Cody, W., Holland, D., Panek, R., Lu, G. and Doherty, A.M., In Kaumaya P.T.P. and Hodges R.S. (eds.) Peptides:Chemistry and Biology (proceedings of the 14th American Peptide symposium), Mayflower Scientific Ltd. U.K., 1996, p. 414.
5. Eaton, S.R., Cody, W.L., Panek, R.L., Lu, G.H. Dahring, T.K., and Doherty, A.M., In 25th National Medkoinal Chemistry Symposium, Ann Arbor, MI USA, 1996.
6. Leonard, D.M., Eaton, S.R., Sawyer, T.K., Doherty, A.M., Boton, G., Hodges, J., Schoten, J., Gowen, R. and Leopold, J.S., In Proceeding of the 23ed European Peptide Symposium, Braga, Portugal, September1994, ESCOM, Leiden,1995, p. 658.
7. Shuler, K.R., Leonard, D.M., Bolton, G.L., Scholten, J., Sebolt-Lecpold, J.S. and Doherty, A.M.,In 21th ACS National meeting, New Orleans, LA March 1996, MEDI #86.
8. Boron, G., Creswell, M., Hodges, J., Gowan, R., Scholten, J. and Sebolt-Leopoki, J., In 25th National Medicinal Chemistry Symposium, Ann Arbor, MI USA, 1996.

Facile new method for preparation of optically active protected proline

Jun-Ichi Yamaguchi and Masaaki Ueki
Department of Applied Chemistry, Faculty of Science, University of Tokyo, 1-3 Kagurazaka, Shinjuku-ku, Tokyo 162, Japan

Introduction

Among the common α-amino acids, proline is useful as a chiral source in asymmetric synthesis. Many methods including asymmetric synthesis of proline has been reported [1]. However, development of a new and convenient method for preparation of optically active proline derivatives is still required. We now report that treatment of L- or D-*N*-protected 2-amino-5-bromopentanoic acid ester (**1**), which was prepared from protected glutamic acid, with sodium hydride gave the corresponding L- or D-proline derivative (**2**) in high yield respectively (Scheme 1).

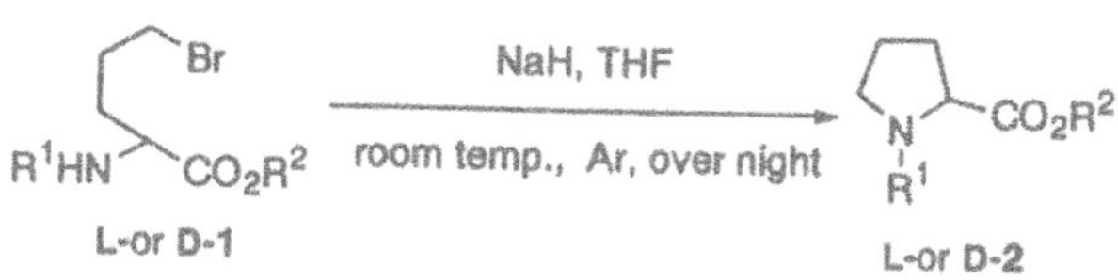

The reaction of 2-amino-4-bromobutyric acid derivative, which was easily prepared from aspartic acid, proceeded under the same conditions to give protected 1-aminocyclopropane-1-carboxylic acid in 74% yield.

Protected L-pipecolic acid was also obtained by the reaction of protected L-2-amino-6-bromohexanoic acid with sodium hydride.

Results and Discussion

Various *N*-protected glutamic acid α-alkyl esters were converted into **1**. The mixed anhydrides prepared from protected glutamic acid and isobutyl chloroformate in the presence of *N*-methylmorpholine (NMM) were reduced with sodium borohydride in tetrahydrofuran (THF) to the corresponding alcohols in 79-95% yields [2].Reacting alcohol with carbon tetrabromide and triphenylphosphine in CH_2Cl_2 at room temperature gave **1** in 76-94% yield [3].

The reaction of *N*-benzyloxycarbonyl (Z) 2-amino-5-bromopentanoic acid *t*-butyl ester with sodium hydride in THF at room temperature conveniently gave *N*-Z-proline *t*-butyl ester in 80% yield. Its structure was supported by ^{1}H NMR and mass spectra, melting point (43-45°C; ref. 44-45°C [4]), and amino acid analysis after deprotection by Hbr/AcOH. The cyclization of **1b** also proceeded to give **2b** in 80% yield. In contrast no reaction was found when **1b** was treated by other reagents (triethylamine, 1,7-diazabicyclo[5.4.0]undec-7-ene

*Table 1. Transformation of **1** into protected proline (**2**)*

Substrate	R1	R2	Product	Yield %
1a	Z	*t*-Bu	**2a**	80
1b	Z	Me	**2b**	80
1c	Z	CH_2Ph	**2c**	83
1d	Fmoc	Me	**2d**	26[a]
1e[b]	Z	CH_2Ph	**2e**	83

[a] The starting material was recovered in 21% yield.
[b] D-Glutamic acid was used as starting material.

(DBU), and potassium *t*-butoxide) instead of sodium hydride under the same conditions. Since the 9-fluorenylmethyloxycarbonyl (Fmoc) group was removed partially in treatment with sodium hydride, *N*-Fmoc-proline derivative was obtained in low yield.

L- and D-prolines were obtained by deprotection of L- and D-protected prolines in 87 and 82% yields, respectively. Adamson *et al.* reported the use of 1-fluoro-2,4-dinitrophenyl-5-alanine amide (Marfey's reagent) to determine the D, L ratio of some amino acids [5]. The coupling product of proline with Marfey's reagent gave diastereomers separable by reverse-phase HPLC. Peaks corresponding to Marfey's derivatives of L- and D-prolines were identified by comparison with D, L-proline-Marfey's reagent standard run under identical conditions. From these experiments, we found that little racemization (<3%) was detected in both cases

Based on the analogous experiment, it was assumed that treatment of 2-amino-4-bromobutyric acid derivative (**3**) with sodium hydride may give the corresponding 4-membered ring compound. However, it was found that the reaction of **3** with sodium hydride proceeded to give 1-amino-cyclopropane-1-carboxylic acid derivative (**4**) in 74% yield (Scheme 2). Its structure was confirmed by ^{1}HNMR spectral data [6].

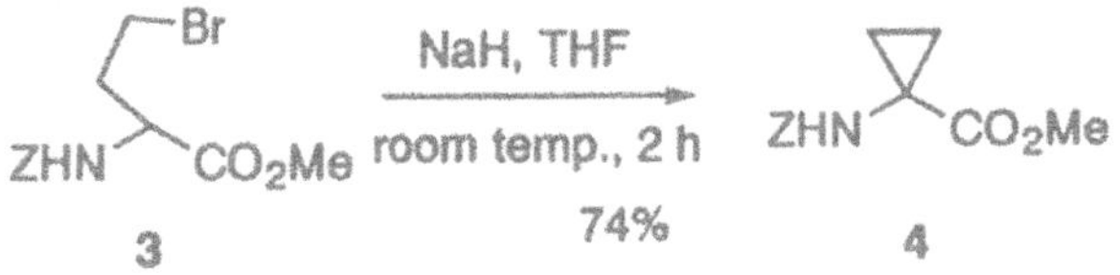

We next synthesized the optically active precursor leading to expensive L-pipecolic acid derivative. Protected 2-amino-6-bromohexanoic acid (**6**) was prepared from protected aspartic acid *via* protected homoglutamic acid (**5**).

The cyclization of **6** under the same reaction conditions gave *N*-Boc-L-pipecolic acid *t*-butyl ester (7) in excellent yield (Scheme 3).

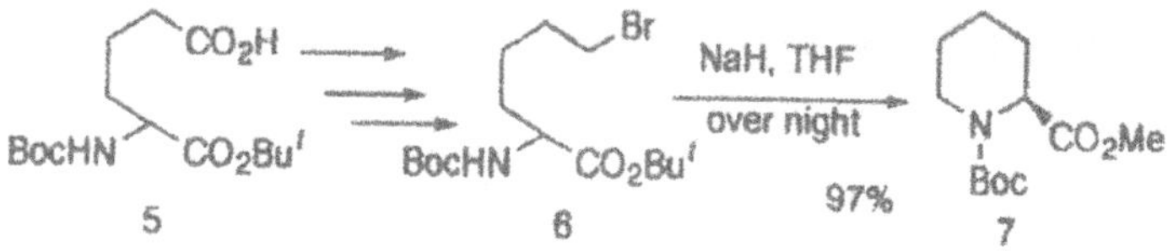

References

1. For example, Buyle, R. Chem. Ind., (1966) 380; Iwata, M. and Kuzuhara, H., Chem. Lett., (1985) 1941.
2. Ishizumi, K., Koga, K., and Ymamada, S., Chem. Pharm Bull., 16 (1968) 492.
3. Kocienski, P. J., Cernigliaro, G., and Feldstein, G., J. Org. Chem., 42 (1977) 353.
4. Anderson, G. W. and Callaham F. M., J. Am. Chem. Soc., 82 (1960) 3359.
5. Adamson, J. G., Hoang, T., Crivici, A., and Lajoie, G. A., Anal. Biochem., 202 (1992) 210.
6. mp 60-62°C. ^{1}H NMR(60MHz) TMS(CDCl3) d=1.0-1.7(m, 4H), 3.58(s, 3H),5.07(s, 2H), 5.40(brs, 1H, this signal disappeared by adding MeOH-*d*4), 7.25(s, 5H)

A pragmatic way to cleave acid-resistant resin

De-Xin Wang and Lei Huang
Institute of Materia Medica, Chinese Academy of Medical Sciences, Beijing 100050, China

Introduction

Nitroficated handle structure between peptide chain and resin carrier was conventionally used as photolabile linkage in solid-phase peptide synthesis before [1,2]. However, it has never been commonly used like other handle structures, such as $-CH_2(C_6H_4)-$, $-CH_2(C_6H_4)-OCH_2(C_6H_4)-$, MBHA, PAM, ... etc., just because of the apparatus restriction in photolytic cleaving experiment. The introduction of $-NO_2$ group into phenylene ring of handle structure can dramatically reduce the electron density from the carbonyl group at benzyl ester moiety. As a result, the ester bonds in nitro-resin should be much more acid resistant than the same bonds in non-nitroficated resin. Peptide should be released more easily from nitro-resin than from other resins by saponification.

Results and Discussion

BNR carrier, nitro-resin, was prepared and used as solid support for the synthesis of peptides (Gly-Asp-Ala-Lys-Gly)$_2$>Lys-OH(**1**) and Thr-Arg-Ser-Ala-Phe-NH_2(**2**). Classical chloromethyl resin (CR) was also used for the synthesis of peptide(**1**) as a criterion of evaluating product yields between different resin supported syntheses. All coupling cycles were carried out by means of Boc-Bzl strategy. DCC/DIEA/HOBt/DCM was the coupling reagent [3] and DDSi/phenyl/DCM was applied to remove Boc groups [4] in each cycle. Before saponification or aminolysis, Hi-TFMSA procedure [5] was performed to remove side-chain protecting groups (SPG).

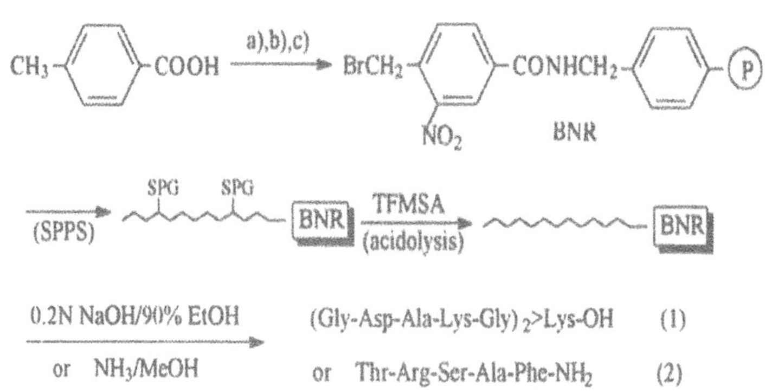

a) NBS, b) HNO_3, c) DCC/H_2NCH_2-(C_6H_4)-polystyrene.

Table 1. Release and purities of peptides

Peptide-resin	by TFMSA	by 0.2N NaOH	purity[a]	Total yield
(1)-BNR	0	92.1%	91.6%	77.3%
(2)-BNR	0	94.8%	90.2%	76.1%
(3)-CR	90.4%	4.8%	62.6%	50.3%

[a] by quantitative analysis of amino acids

The results shown in Table1 indicated that the benefit of using a combination of cleavage of acidolysis-saponification (or aminolysis) as an ogonal protocol. Introduction of nitro group to the chloromethyl resin did save the integrity of peptidyl-resin from handle break under strong acidic condition and promote saponification to completeness. In addition, other advantages stated below are also attractive:

1. BNR handle was durable to de-Boc reagent. So there was little possibility to lose peptide chain from resin in the coupling cycles.
2. It would be possible to perform more drastic de-Boc conditions, in case of assembling a peptide with "difficult sequence", to ensure complete deprotection-coupling cycles.
3. Even mild conditions of acidolysis (Low-TFMSA procedure [5]), to remove the SPGs.

Could simultaneously cleave peptide from classical CR resin (Table.1). Free peptide, in this case, would be mixed with a lot of SPGs and scavengers in the crude product. On the contrary, the successive cleavage-saponification or aminolysis would proceed in a "pure circumstance",with naked peptidyl-resin only but no more SPGs and no scavengers. As a result, the purity of crude product could be improved.

References

1. Rich, D. H. and Gurwara, S. K., J. Am. Chem. Soc., 97 (1975) 1575.
2. Giralt, E., Tetrahedron, 38 (1982) 1193.
3. Beyermann, M., Int. J. Pept. Prot. Res., 37 (1991) 252.
4. Wang, D.X., Chin. J. Med.Chem., (1991) 60.
5. Tam, J.P., Heath,W.F. and Merrifield, R.B., J.Am. Chem. Soc., 108 (1986) 5242.

Phosphonamidate based inhibitors of angiotensin - converting enzyme (ACE)

He-Ru Chen[a] and Jie-Cheng Xu[b]
[a]*Department of Chemistry, Shantou University, Shantou 515063, China*
[b]*Shanghai Institute of Organic Chemistry, Chinese Academy of Sciences, Shanghai 200032, China*

Introduction

The development of the first clinically available orally active ACE (angiotensin - converting enzyme) inhibitor, captopril [1,2] by Squibb group in 1977 started a new approach to antihypertension therapy. Reviewing the history of the captopril development, one of the key ideas in this process was a hypothesis that ACE is related structurally and mechanistically to carboxypeptidase A. This allowed the Squibb group to develop a hypothetical model of the ACE active site which led to the design of the first non-peptide inhibitor [3].

The isolation and structure determination of phosphoramidon and its potent inhibition of the zinc metalloenzyme thermolysin were reported in 1973 [4,5]. X-ray crystallographic studies on its E-I complex with thermolysin established its active site binding interaction and led to the proposal that its hydroxyphosphinyl moiety binding to Zn^{++} resembles the presumed transition state for amide hydrolysis [6]. It is not surprising then that the hydroxyphosphinyl group has been extensively studied for possible applicability to the design of ACE inhibitors.

Results and Discussion

A series of phosphonotripeptide inhibitors were designed and synthesized by conventional method as described previously [7]. A column packed with Nucleosil (43 ~ 60μm) was used to separate the epimeric phosphonotripeptides. All peptides were confirmed by MS, elemental analysis, ^{1}H and ^{31}P-NMR.

Kinetic properties of inhibition of the synthesized phosphonotripeptides toward ACE have been studied by applying the simple and sensitive method developed by our group [8]. The results are shown in Table 1. From Table 1, it can be concluded that the inhibitive types of synthesized phosphonotripeptides are all competitive. The inhibition activities of compounds 1-2, 1-6, 1-9 toward ACE are roughly equipotent to captopril.

Some interesting points can be drawn from the correlation of structure - activity.In three pairs of epimers (compounds 1-2/1-2i, 1-6/1-6i, 1-9/1-9i), when the configuration of the epimeric chiral center is R, the inhibition activity is 7 ~ 11 times higher than the S epimer. This might be due to the hydrophobic property of S_1 subsite in ACE. As for the R - epimer, the lipophilic group R_1 could be more adjacent to S_1 subsite and therefore it will enhance the affinity between inhibitor and enzyme. The hydrophobic S_1 subsite can also be shown by the 30 - fold lower activity as in compound 1-3 when the phenylmethyl group in 1-2 is replaced by a more hydrophilic group p-hydroxylphenylmethyl.

Table 1. Inhibition properties of phosphonotripeptides

Comp.	Structure	Inhibitive types	I_{50}(nm)
*Captopril	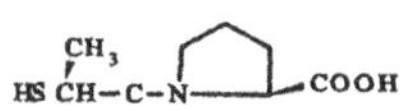	Competitive	1.8
1–1	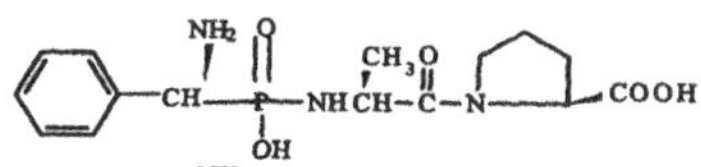	~	58
1–2	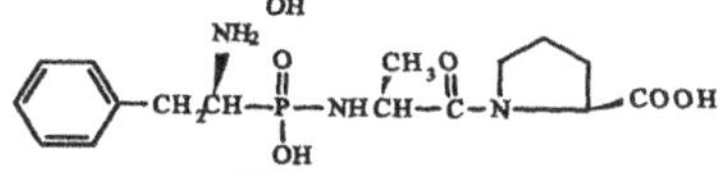	~	4.2
1–2i	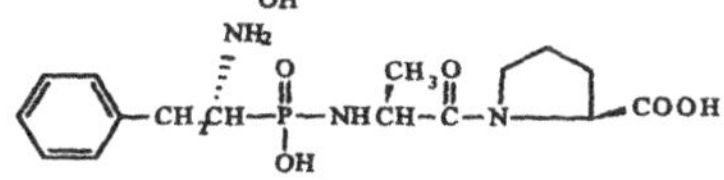	~	32
1–3	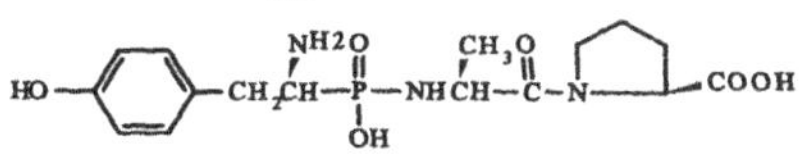	~	130
1–4	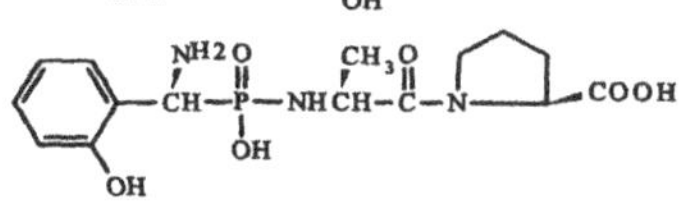	~	180
1–5	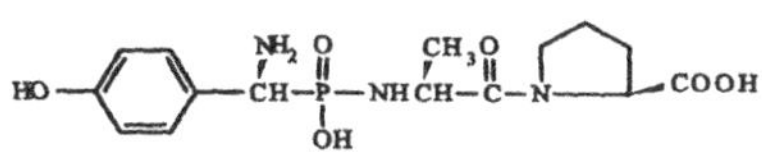	~	79
1–6	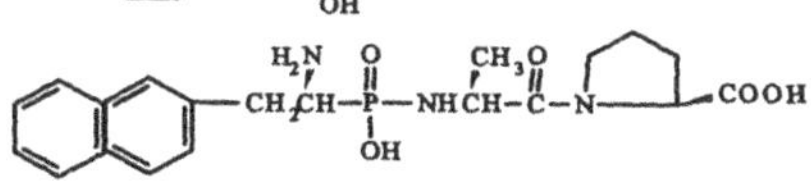	~	7.2
1–6i	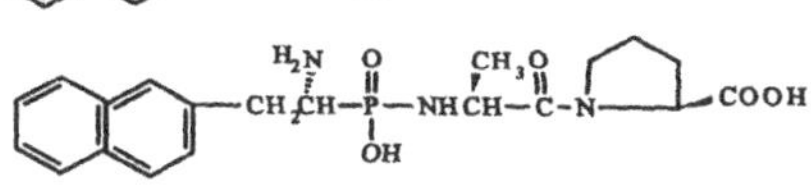	~	52
1–7	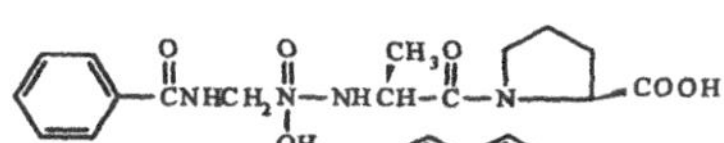	~	680
1–8	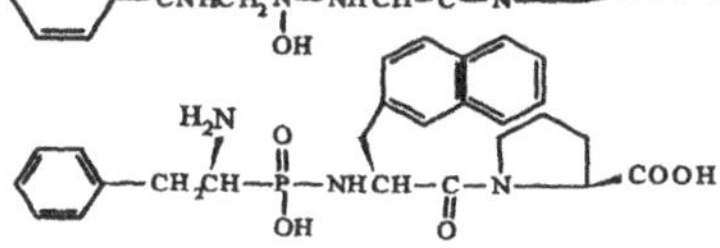	~	5.4
1–9	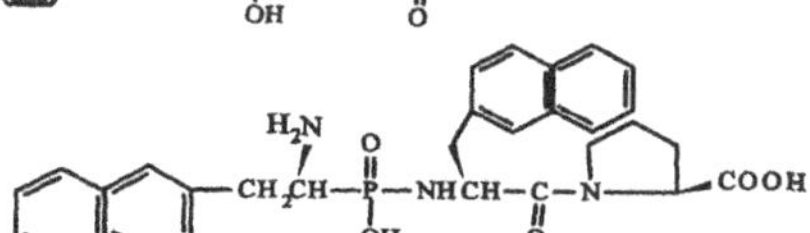	~	8.7
1–9i	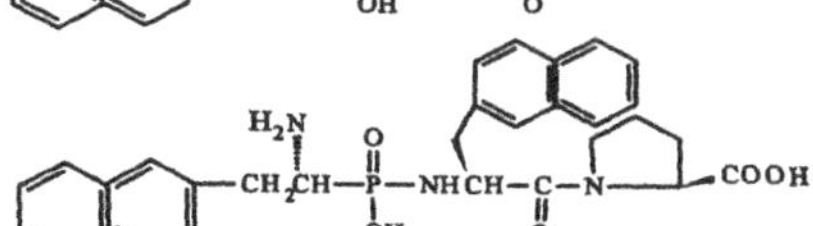	~	102
1–10	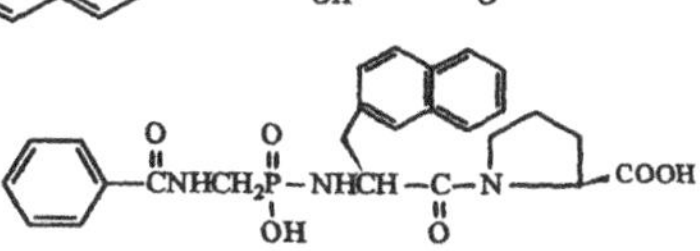	~	640

* purchased from Sigma Chem. Co.

Benzamido derivatives (compounds 1-7, 1-10) without alkyl group (R_1= H) caused eminent decrease in activity. It has been pointed out by Patchett [9] that the P_1 site of inhibitor which fits well to the S_1 subsite in enzyme is a key factor in effective inhibition.

Replacement of the phenyl substituents in compound 1-2 or 1-8 with naphthyl in compound 1-6 or 1-9, resulted in a slight decrease of activity. It is assumed that larger group in P_1 site cannot match S_1 subsite well. Comparing of compound 1-1 with 1-2, it is that almost 16 - fold decrease in activity caused by only one methylene moiety difference in P_1 site. Similarity between 1-3 and 1-5 with 1-2 and 1-1 can also be found, but compound 1-5 is more potent than 1-3.

The activity enhancing properties of S_1' subsite in ACE probably have a conformation origin as suggested for captopril [2],that is, the substituent in this position must be of the L-configuration and hydrophobic groups larger than CH_3 do not lead to activity enhancement in inhibitors. The latter point is supported by our experiments (compounds 1-2/ 1-8, 1-6/1-9, 1-6i/1-9i).

Acknowledgment

The authors thank the National Natural Science Foundation of China for financial support.

References

1. Ondetti M.A., Rubin, B. and Cushman, D.W., Science, 196 (1977) 441.
2. Cushman, M.A., Cheung, H.S., Sabo, E.F.and Ondetti, M.A., Biochemistry 16 (1977) 5485.
3. Edward, W., Petrillo, Jr. and Ondetti, M.A., Med. Res. Rev., 2 (1982) 1.
4. Umezawa, S., Tatsuta, K., Izawa, O., and Tsuchiya, T., Tetrahedron Lett., 13 (1972) 97.
5. Suda, H., Aoyagi, T., Takeuchi, T., and Umezawa, H., Antibiot. J., 26 (1973) 621.
6. Waver, L.H., Kester, W.R., and Matthewa, B.W., Mol. Biol. J., 114 (1977) 119.
7. Chen, H.R. and Xu, J.C., Proceeding of the Chinese Peptide Symposium 1994, p. 51.
8. Chen, H.R. and Xu, J.C., Anal. Biochem., to be published.
9. Patchett, A.A., Harris, E., Tristram, E.W. and Wyvratt, M.T., Nature (London), 28 (1980) 280.

Thionyl chloride-phenol and phosphorus oxychloride-phenol could be used as efficient deprotection agents in peptide synthesis

Zhen-Wei Miao, Yi Pan, Xiang-Qun Li, Xiao-Jie Xu and You-Qi Tang
Department of Chemistry, Peking University, Beijing 100871, China

Introduction

Hydrogen chloride in organic solvent has long been used for removing t-Boc protection in peptide synthesis [1]. However, this method is limited with its tedious operation and low efficiency. Endeavor to an alternative deprotecting reagent has recently been made by Kaiser and Tam and co-workers [2,3] using Me_3SiCl-PhOH in DCM. As an attempt to search for cheaper and more convenient deprotecting agent, $SOCl_2$-PhOH (reagent S) and $POCl_3$-PhOH (reagent P) were designed in this paper with the same criteria as Kaiser/Tam's method. Phenol served as proton donor and carbonium ion scavenger, however, Me_3SiCl was replaced by $SOCl_2$ or $POCl_3$. Reagents S and P were first tested in model experiments to investigate their deprotecting efficiency, and then were employed in solid phase peptide synthesis (SPPS).

Results and Discussion

In model experiments, 50mg Boc-Lys(Z)-Gly-OBzl was dissolved in 5ml 2-6M PhOH/DCM, followed by adding $SOCl_2$/DCM or $POCl_3$ with equal molar ratio. The process of deprotection was monitored by thin layer chromatography or HPLC. The results demonstrated that the mechanism of deprotection by reagent S was different from that by reagent P. Reagent S was in S_N1 mechanism, whereas reagent P in S_N2 mechanism. When freshly prepared 1-3M reagent S (mixing 2-6M PhOH/DCM and corresponding $SOCl_2$/DCM in V:V=1:1) was used, the t-Boc group could be completely removed within 5 minutes without affecting the integrity of the benzyl-based protecting groups (Bzl or Z). If the prepared reagent S stood for a period before use, its deprotecting ability decreased dramatically, much longer time was taken to give a full removal of t-Boc. The effective proton concentration of reagent S decreased due to the continuous escape of hydrogen chloride at the same time. When nonpolar solvents such as tetrahydrofuran and ethyl acetate were used to overcome this problem, the deprotecting ability of reagent S was unexpectedly reduced, since the hydrogen-bond formed with the solvents weakened the proton strength. Evidence above demonstrated that the deprotection process by reagent S is an S_N1 mechanism involving the participation of the produced hydrogen chloride, that is, the effective proton concentration and strength were determined the deprotecting efficiency. In contrast to reagent S, reagent P in DCM took about 10 minutes to remove t-Boc completely. Its deprotecting ability would not decrease and no HCl gas was found while it stood even for several months. The polar solvents had little influence on its deprotecting ability. Combined with the kinetic ^1H and ^{31}P NMR studies, it could be concluded that the deprotecting process of reagent P is an S_N2 mechanism, similar to Me_3SiCl-PhOH [3]. In addition, when Boc-Tyr(Bzl)-OH was employed to substitute Boc-Lys(Z)-Gly-OBzl in model experiments, similar results were achieved. When PCl_3 were replaced $SOCl_2$, the deprotecting reaction adopted the same mechanism as reagent S, but was found to be more vigorously. Also, it is worthy to point out that Z and Bzl were stable enough even for 72 hours under the deprotection condition both by reagent S and P.

Based on the results of model experiments, we chose 2.0M reagent P and freshly prepared 1.5M reagent S as deprotecting agents in the synthesis of IA-16, H_2N-S(EEIAKKI)$_2$G-$CONH_2$ (a *De Novo* designed peptide with 16 amino acid residues), and compared it with 33% TFA/DCM in a paralleled control synthesis. The peptide was synthesized by stepwise SPPS using t-Boc/Bzl chemistry [4]. Deprotection was done by reagent S or P first for 5 minutes and then for 30-40 minutes. As expected, both reagents S and P exhibited high efficiency, the complete deprotection was always obtained by Kaiser test [5], in contrast, 33%TFA/DCM in controlled experiment sometimes did not give complete deprotection. After peptide chain assembly was finished, the peptide was cleaved from MBHA resin by anhydrous HF and then purified by HPLC.

Figure 1 shows the HPLC profile of crude IA-16 using reagent S as deprotecting agent. Peak 3 was the target peptide as confirmed by FAB-MS $(M+H)^+$: found 1785, calculated 1785) and amino acid composition analysis. Peaks 1 and 2 could be the additives used in cleavage procedure, and disappeared after lyophilization. 15mg pure products were obtained from 150mg peptide resin with this method. In controlled experiment, the HPLC

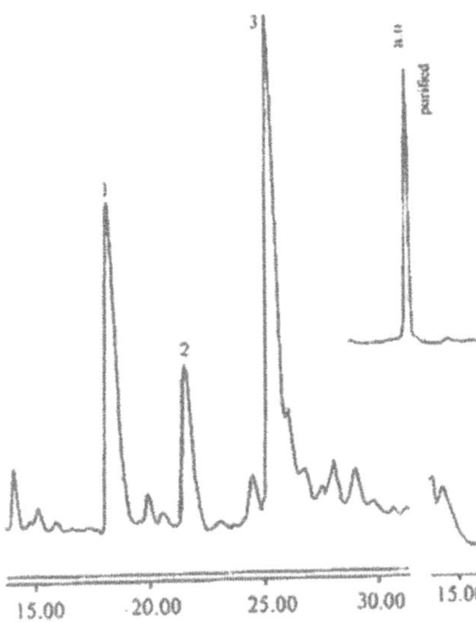

Fig. 1. HPLC Profile of Crude IA-16. Column: Waters C18 (7.8×250mm), 2.4ml/min Gradient: 30-45% acetonitrile/30min, in 0.1%TFA

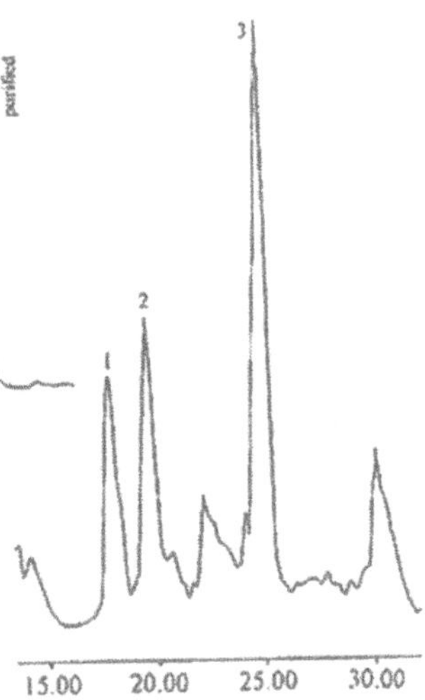

Fig. 2. HPLC Profile of Crude IA-16. Column: Waters C18 (7.8x250mm), 2.4ml/min Gradient: 25-50% acetonitrile/30min, in 0.1%TFA

profile was more complicated (not shown here) and only 11mg of purified peptide was obtained. This result revealed that 1.5M reagent S in DCM as deprotecting reagent gave a better product both in purity and yield than 33% TFA/DCM. However, this reagent seemed to be unsuitable for automatic SPPS. In the course of deprotection, a lot of gas including HCl, CO_2 and isobutylene was produced in a short time. It would be dangerous unless the reaction vessel is pressure resistant. Therefore, reagent S is recommended to be used in solution method and manual SPPS only. Similarly, when 1.5M reagent P was used as deprotecting agent, the result was satisfactory (Fig. 2). Reagent P was slightly advantageous over reagent S. It could give higher yield (19mg pure peptide could be obtained from 150mg peptide resin). Moreover, it could be used in automatic SPPS for its smooth deprotecting procedure.

References

1. Schnabel, E., Ann. Chem., 702 (1967) 188.
2. Kaiser, E., Tam, J.P., Kubiak,T.M. and Merrifield, R.B., Tetra. Lett. 29 (1988) 303.
3. Kaiser, E., Picart, F., Kubiak, T., Tam, J.P. and Merrifield, R.B., J. Org. Chem. 58 (1993) 5167.
4. Stewart, J.M. and Young, J.D. Solid phase peptide Synthesis, Pierce Chemical, Rockfold,IL, 1984.
5. Kaiser, E., Bossinger, C.D. and Cook, P.I., Anal. Biochem. 34 (1970) 595.

Solid phase synthesis of sodium ion channel IS3-4 segment simultaneously using Boc and Fmoc chemistry

Zhen-Wei Miao, Yun Jiang, Xiao-Jie Xu and You-Qi Tang
Department of Chemistry, Peking University, Beijing 100871, China

Introduction

Both IS3 and IS4 are two of the six transmembrane segments of the rat brain sodium ion channel [1,2]. Interestingly, individual IS3 can mediate the ionic conductance in planar lipid bilayers lack of selectivity between sodium and potassium ions [3,4]. On the other hand, IS4, a highly conserved sequence with regular alternation of one positive-charged and two hydrophobic residues, is predicted to play an important role in switching the channel open and close [5,6]. In order to study the interaction between IS3 and IS4, a peptide with 47 residues (IS3-4) was designed by linking IS3 and IS4 with a connecting sequence GRG, and synthesized by a combined strategy, using Boc and Fmoc chemistry.

IS3-4:	DPWNWLDTFVITFAYVTEFVDL	GRG	SALRTFRVLRALKTISVIPGLK
	IS3	linker	IS4

Results and Discussion

Boc/Bzl and Fmoc/Bu^{t} strategies are often used in solid phase peptide synthesis (SPPS) [7,8]. Usually, one chemistry is used in one procedure for one peptide. However, such a protocol could not provide a satisfactory result for the synthesis of IS3-4. In our initial procedure, Boc chemistry was employed alone, however, the resulting product was very complicated. Particularly, it was extremely hydrophobic and was difficult to be dissolved in acetonitrile/water mixture. The purification of the crude product by HPLC consequently became arduous. The C-terminal 25 residues including the IS4 and the linker sequence could be obtained in high purity and high yield by Boc chemistry. It could thus be deduced that the synthesis problem resulted from the IS3 segment, because it contains several Trp, Tyr, Asp and Asn residues, which were usually subjected to side reactions during deprotection by 33% TFA/DCM. Therefore, in our improved procedure, the Fmoc/Bu^{t} chemistry was introduced instead of the original Boc/Bzl chemistry for the assembly of the remaining 22 residues. The final peptide was cleaved from MBHA resin by anhydrous HF : anisole : Trp = 18 : 2 : 1. As expected, this procedure turned out to be much better than the initial one. The crude peptide could be dissolved in 40% acetonitrile/ water, and then was purified by RP-HPLC using C18 column (Fig.1) on Gilson UniPoint HPLC System. The HPLC profile exhibited only two major peaks (p1, p2).

The TOF-MS spectrum confirmed that p2 was the desired product (calculated: 5384.28, found: 5384.5). The result of amino acid composition further supported that p2 was the expected IS3-4: Ala 3.14(3), Arg4.09(4), Asp 3.93(1Asn,3Asp), Glu 1.04(1), Gly 3.26(3), Ile 2.73(3), Leu 6.29(6), Lys 2.28(2), Phe 4.00(4), Ser 2.07(2), Thr 4.84(5), Tyr 0.80(1), Val 4.81(5). Thus, it could be concluded that the combined strategy using Boc and

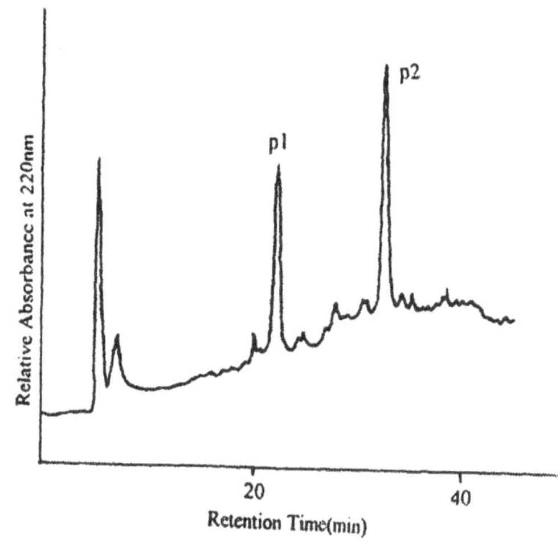

Fig. 1. HPLC Profile of IS3-4.
Column: Cosmosil C18(7.8x250mm), Flow rate: 2.4ml/min
Gradient: 50-90% acetonitrile/30mins in 0.1% TFA aqueous.

Fmoc chemistry in an integrated fashion could provide a satisfactory result for large hydrophobic peptide synthesis.

References

1. Casadei, J.M., J. Biol. Chem., 261 (1986) 4318.
2. Caterall, W.A., Science, 202 (1988) 50.
3. Labara, P. Lindstron, J. and Montal, M., J. Gen. Physiol., 83 (1984) 473.
4. Labara, P. Rice and J.A. Fredkin, D.R., Montal, M. Biophys. J., 47 (1985) 469.
5. Guy, H. and Seetharamulo, P., Proc. Natl. Acad. Sci. USA, 83 (1986) 508.
6. Kosower, E.M., FEBS Lett., 234 (1985) 1982.
7. Stewart, J.M. and Young, J.D. Solid Phase Peptide Synthesis, Pierce Chemical, Rockford, IL, 1984.
8. Fields,G.B. and Noble, R.L. Int. J. Pept. Prot. Res. 35 (1990) 161.

Influence of pH and water content on enzymatic peptide synthesis in organic solvents

Guo-Wen Xing, Gui-Ling Tian, Ying Lu* and Yun-Hua Ye
Department of Chemistry, Peking University, Beijing 100871, China

Introduction

Enzymatic catalyzed reactions in organic solvents with a small amount of water content possess many advantages, such as great stability of enzymes in organic solvents etc.. We focused our attention on the influence of pH and water content on enzymatic peptide synthesis in organic media. N-P-L-Asp-XaaOR (P=Z or For, XaaOR=L-PheOMe or L-AlaOcHex, cHex=cyclohexyl), the precursors of aspartame [1] and a new sweetener(L-Asp-L-AlaOcHex) [2, 3], and N-Z-L-TyrGlyGlyOEt, a fragment of Leu-enkephaline as model peptides were synthesized by enzymes in organic solvents [4]. All the peptides were identified by MS, elemental analysis and optical rotation and they were identical with those peptides synthesized by chemical method.

Results and Discussion

1. Z-L-Asp was reacted by L-PheOMe with thermolysin in *tert*-amyl alcohol. The optimum pH was 8 and water content was 6-8%(V/V) (Fig. 1, 2). Under these optimum conditions, when the molar ratio of carboxyl component and amino component was 1:1, 1:2, 1:3, the isolated yield was 41.3, 72.5, 82.7% respectively. In the synthesis of Z-L-Asp-L-AlaOcHex, the results indicated that the optimum pH was 9 and water content was 6%(V/V) (Fig. 1, 2). Under these conditions, when molar ratio was 1:1, 1:2, 1:3, the isolated yield was 11.5, 20.5, 44.1% respectively. From the experimental data we can see that L-PheOMe is a better nucleophilic component than L-AlaOcHex for thermolysin.
2. Under same reaction conditions as Z-L-Asp-L-PheOMe and Z-L-Asp-L-AlaOcHex, the isolated yields of For-L-Asp-L-PheOMe and For-L-Asp-L-AlaOcHex were 47.4% and 6.4% respectively. The lower yields showed that formyl was not a good protecting group for thermolysin reactions in *tert*-amyl alcohol.
3. N-Z-L-TyrOEt was coupled with GlyGlyOEt catalyzed by α-chymotrypsin powder suspended in dichloromethane. The results showed that the optimum water content was found to be 0.15%(V/V) when pH was 10 (Fig. 3), the isolated yield was 71.3%.

From above results we can conclude that the pH and water content in enzymatic peptide synthesis are important. The essential water is quite different from various enzymes in organic media. In our study, we found that the optimum water content in synthesis of Z-L-Asp-L-AlaOcHex was 40 times that of N-Z-L-TyrGlyGlyOEt. The physical constants of peptides synthesized by thermolysin and α-chymotrypsin were listed in Table 1.

* Visiting teacher of Peking University from Harbin Research and Development Center of Pharmaceutical Engineering and Technology.

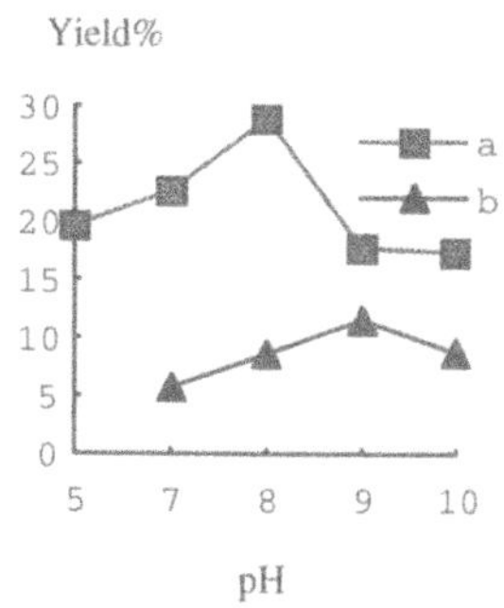

Fig. 1. Effect of pH on the isolated yield catalyzed by thermolysin in tert-amyl alcohol. a: Z-AspPheOMe b: Z-AspAlaOcHex

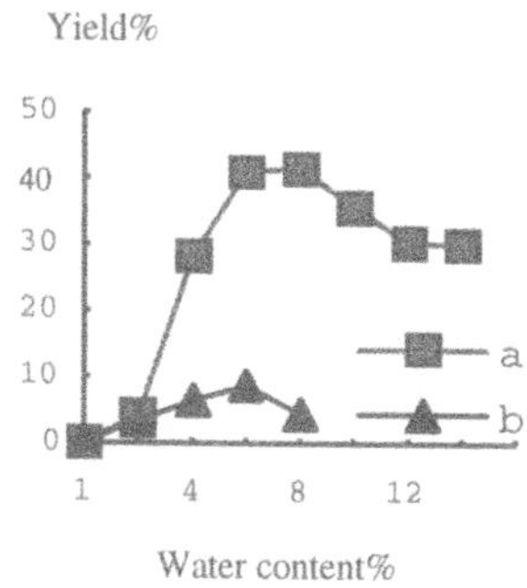

Fig. 2. Effect of water content on the isolated yield catalyzed by thermolysin in tert-amyl alcohol. a: Z-AspPheOMe b: Z-AspAlaOcHex

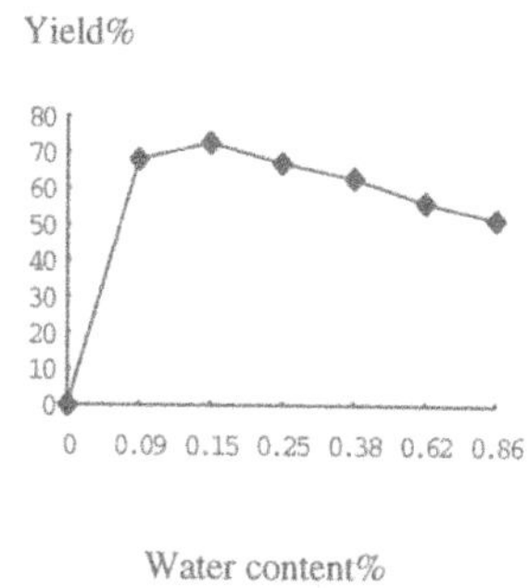

Fig. 3. Effect of water content on the isolated yield of Z-TyrGlyGlyOEt catalyzed by α-chymotrypsin in dichloromethane.

Table 1. Physical constants of peptides synthesized by thermolysin and α-chymotrypsin

No.	Enzyme	Product[a]	Solvent[b]	Yield%	m.p.(°C)	$[\alpha]_D^t$ (c, sol.)[c]
1	thermolysin	Z-AspPheOMe	*tert*-amyl alcohol	82.7	120-122	-14.8(0.5, MeOH)
2	thermolysin	Z-AspAlaOcHex	*tert*-amyl alcohol	44.1	119-121	-31.6(0.5, MeOH)
3	thermolysin	For-AspPheOMe	*tert*-amyl alcohol	47.4	138-140	-37.3(0.5, MeOH)
4	thermolysin	For-AspAlaOcHex	*tert*-amyl alcohol	6.4	144-147	-54.2(0.25,MeOH)
5	α-Chy[d]	Z-TyrGlyGlyOEt	dichloromethane	71.3	165-166	+2.9(2, HAc)

[a] [acyl donor]/[Nucleophile] in No. 1-4 was 1:3; in No. 5 was 1:1.
[b] Water content(V/V) in No. 1, 3 was 8%; in No. 2, 4 was 6%; in No. 3 was 0.15%.
[c] t: measured temperature, in No. 1- 4 was 25°C; in No. 5 was 20°C.
[d] α-Chy is the abbreviation of α-chymotrypsin.

Acknowledgment

The authors thank the National Natural Science Foundation of China for financial support (No. 29472033).

References

1. Mazur, R.H., Schlatter, J.M. and Goldkamp, A.H., J. Am. Chem. Soc., 91 (1969) 2684.
2. Zeng, G.Z., Chen, J.T., He, H.Z., Wang, Z.Q. and Yan, J.S., J. Agric. Food Chem., 39 (1991) 782.
3. Yuan, Z.F., Zhu, T.S. and Lai, J.Y., Chinese Chem. Lett., 6 (1995) 369.
4. Tian, G.L., Liu, Y., Wang, H. and Ye, Y.H., Chem. J. Chinese Univ., 17 (1996) 55.

Session II
Molecular Diversity-Peptide Libraries

Chairs: Victor J. Hruby
University of Arizona
Tucson, Arizona, U.S.A.

Arnold Satterthwait
The Scripps Research Institute
La Jolla, California, U.S.A.

Zhen-Kai Ding
Chinese Institute of Military Medical Science
Beijing, China

and

Bing-Gen Ru
Peking University
Beijing, China

Application of the one-bead one-compound combinatorial library method in protein tyrosine kinase and cell surface receptor research

K.S. Lam, Q. Lou, J. Wu, M. Leftwich, R.T. McKay, L. Rychetsky, H. Phan, J. Joe, M.-L. Chen, R. Liu-Stevens, Y. Zhao and S.E. Salmon
Arizona Cancer Center, Department of Medicine, University of Arizona, Tucson, AZ 85724, U.S.A.

Introduction

In the past six years, we have successfully applied the "one-bead one-compound" combinatorial library method to many molecular targets. The "one-bead one-compound" concept is based on our recognition that when a "split-synthesis" method is used in the generation of a compound-bead library, each bead displays one unique chemical compound although there are over 10^{13} copies of the same compound on each bead (120μm diameter) [1-3]. Using an enzyme-linked colorimetric on-bead binding assay, we identified specific ligands for monoclonal antibodies, streptavidin, avidin, MHC-class I molecules, proteases, and even small molecules. Recently, we have developed new screening assays for other molecular targets. For example, we can rapidly determine peptide substrate motifs for protein kinases by incubating peptide-libraries with [γ-^{32}P] ATP and a specific protein kinase. The [^{32}P]-labelled peptide-bead can then be localized by autoradiography and subsequently isolated for structure determination [4-6]. We have also developed a new whole cell on-bead binding assay that enables us to discover peptide ligands for various cell surface receptors. In addition, an *in situ* solution phase releasable assay for the discovery of anticancer agents has also been developed.

Results and Discussion

Protein tyrosine kinase

When a random heptapeptide library (XXXXXXX where X = all 20 L-amino acids except cysteine) was incubated with [γ-^{32}P]ATP and p60c-src protein tyrosine kinase (PTK), a peptide bead, YIYGSFK, was identified as an efficient and specific substrate for p60$^{c\text{-}src}$ PTK [5]. Based on an extensive SAR study [7], we determined that a hydrophobic amino acid at position 2 was crucial for the activity as a substrate. We, therefore, synthesized a secondary library (XIYXXXX) to optimize the initial lead. In a limited library screen, we isolated 5 peptides: GIYWHHY, KIYDDYE, EIYEENG, EIYEEYE, and YIYEEED [6]. GIYWHHY was determined to be a more efficient substrate with a K_m of approximately 20μM. Based on the primary structure of GIYWHHY, we designed and synthesized several pseudosubstrate-based peptide inhibitors for p60$^{c\text{-}src}$ PTK (Table 1). Some of these peptide inhibitors are very potent with an IC_{50} at the low micromolar range. Songyang et al. [8] reported the discovery of an efficient peptide substrate (EEIYGEFF) for p60$^{c\text{-}src}$ PTK using a different combinatorial peptide library approach. Their peptide library was phosphorylated in solution and the phosphorylated peptides were isolated by affinity

Table 1. Pseudosubstrate-based Peptide Inhibitors for p60$^{c\text{-}src}$ PTK Derived from GIYWHHY (6) and EEIYGEFF (8)

Peptide	IC_{50} µM*
GI(2-Nal)WHH(2-nal)	3.9
GI(2-nal)WHH(2-nal)	4.0
GI(2-Nal)WHH(2-Nal)	4.2
GI(2-Nal)WHH(Aba)	7.0
GI(2-Nal)WHHY	23
GI(2-Nal)WHH	26
GlyWHH	50
EEI(2-Nal)GEFF	143
EEIyGEFF	867

2-Nal, (L) 2-naphthylalanine; 2-nal, (D) 2-naphthylalanine; y, D-tyrosine; Aba, aminobenzoic acid. *The values represent the mean of 2-4 independent experiments. The substrate used in these experiments was YIYGSFK (55 µM).

Table 2. Branched Chimeric Peptide Inhibitors for p60$^{c\text{-}src}$ PTK

Peptide	IC_{50} µM
YI(2-Nal)GKFK | HHW	0.6
YIYGKFK | HHW	2.2

The values represent the mean of two independent experiments. The substrate used in these experiments was YIYGSFK (55µM).

chromatography and the retrieved peptides were microsequenced collectively. Pseudosubstrate based peptide inhibitors which are derived from EEIYGEFF exhibited only weak inhibitory activity towards p60C$^{c\text{-}src}$ PTK (Table 1).

Both YIYGSFK (Km ~ 55 µM) and GIYWHHY (Km ~ 20µM) are relatively efficient and specific substrates for p60$^{c\text{-}src}$ PTK. However, the only homology between these two peptides is IY. This suggests that the enzyme active site interacts with these two peptides differently. We hypothesize that "chimeric" derivatives of these two peptides may be more potent as inhibitors for p60$^{c\text{-}src}$ PTK. The two branched chimeric peptides shown in Table 2 were found to have strong inhibitory activity for p60$^{c\text{-}src}$ PTK. YI(2-Nal)GK(HHW)FK is the most potent inhibitor with an IC_{50} of 0.6µM. Interestingly for YIYGK(HHW)FK, although Tyr-3 is a potential phosphorylation site, this chimeric peptide could not be phosphorylated by p60$^{c\text{-}src}$ PTK. Instead it is a relatively potent inhibitor for the enzyme with an IC_{50} of 2.2 µM. The fact that these chimeric branched peptides are excellent inhibitors suggests that the peptide pocket of the p60$^{c\text{-}src}$ PTK enzyme active site is relatively big and can accommodate more than a short linear peptide.

Whole cell on-bead binding assay

We have previously shown that intact lymphoma cells can rosette around idiotype-specific peptide-beads [9]. Recently, we have applied this whole cell on-bead binding assay to screen random peptide libraries for cell surface receptor specific peptides. We first mixed intact cells (human lymphoma, or myeloma cell lines) with a thoroughly washed random peptide-bead library in culture medium. After incubation at 37°C for 2 hours, the free cells were removed and the bead-library was transferred to a Petri dish and inspected under a dissecting microscope. Beads coated with a monolayer of intact cells were isolated, treated with 8M guanidine HCl (pH 1.0) and prepared for microsequencing. With this methodology we succeeded in isolating cell surface binding peptides from both the L-and D-amino acid peptide libraries. Blocking antibodies were then used to reprobe the positive beads for determining the identity of the cell surface receptor to which these peptides bind.

In situ solution phase releasable assay

We previously disclosed [3] the potential application of the "one-bead one-compound" library method to discover anti-cancer or anti-microbial compounds using an *in situ* soft agar solution phase releasable assay.Recently, we applied this approach to discover novel anti-cancer agents. In this method the random peptides or small molecules are attached to the solid phase bead via a dual cleavable "Ida linker" [10]. The bead-library was first mixed with a cancer cell line and plated in soft agar. After 48 hours, 3-[4,5-dimethylthiazol-2-yl]-2,5-diphenyltetrazolium bromide (MTT) was added to stain the live cells. Two hours later, a clear zone of inhibition (a halo) was then observed around the beads that released active cytotoxic compounds against the cancer cell line [11]. We have now identified compounds that appear to be specific against certain tumor types. Work is currently underway to confirm these preliminary results.

Conclusion

The "one-bead one-compound" combinatorial library method is extremely versatile and can be used to discover ligands for various molecular targets. Assays can be developed such that a specific biological or physical property can be detected. These assays, whether on-bead or in solution phase can easily be adapted to the "one-bead one-compound" library concept. Thus far, this specific combinatorial library method has proven to be very useful in both basic research and drug discovery.

Acknowledgment

The work is supported by NIH grants CA17094, CA57723, CA23074, and an NSF grant (MCB 9506217). Kit S. Lam is a Scholar of the Leukemia Society of America.

References

1. Lam, K.S., Salmon, S.E., Hersh, E.M., Hruby, V, Kazmierski, W.M. and Knapp, R.J., Nature, 354 (1991) 82.
2. Lebl, M., Krchnak, V., Sepetov, N.F., Seligmann, B., Strop, P., Felder, S. and Lam, K.S., Biopolymers

(Peptide Science), 37 (1995) 177.
3. Lam,K.S. and Salmon,S.E., United States Patent Number 5,510,240. April 23, 1996.
4. Wu, J., Ma, Q.N. and Lam, K.S., Biochemistry, 33 (1994) 14825.
5. Lam, K.S., Wu, J.S. and Lou, Q., Int. J Pept. Prot. Res, 45 (1995) 587.
6. Lou, Q., Leftwich, M. and Lam, K.S., Bioorganic and Medicinal Chemistry, 1996, in press.
7. Lou, Q.,Lou, J., Salmon, S.E and Lam, K. S., Letters in Peptide Science, 2 (1995) 289.
8. Songyang, Z., Carraway III, K.L., Eck, M.J., Harrison, S.C. Feldman, R.A., Mohammodi, M, Slessinge, R. J., Hubbard, S., Mayer, B.J. and Cantley, L.C. Nature, 37 (1995) 536.
9. Lam, K.S., Lou, Q., Zhao, Z.G., Chen, M.L., Smith, J., Pleshko, E. and Salmon, S.E., Biomedical Peptides, Proteins and Nucleic Acids, (1995) 205.
10. Lebl, M., Krchnak,V., Salmon, S.E. and Lam, K.S., Methods in Enzymology, 6 (1994) 381.
11. Salmon, S.E., Liu-Stevens, R.H., Zhao, Y., Lebl, M., Krchnak, V., Wertman, K., Sepetov, N. and Lam, K.S., Molecular Diversity, 2 (1996) 57.

Libraries of synthetic glycopeptides in the characterization of the T cell response to tumor associated mucin antigens

M. Meldal[a], E. Meinjohanns[a], K. Frische[a], T. Jensen[b], P. Hansen[b], O. Werdelin[b], L. Galli-Stampino[c], S. Mouritsen[c] and K. Bock[a]
[a] *Carlsberg Laboratory, DK-2500 Valby, Denmark*
[b] *University of Copenhagen, DK-2200 N Copenhagen, Denmark*
[c] *M&E, DK-2100 Copenhagen, Denmark*

Introduction

It has been well established that T cell stimulation is mediated by the presentation of peptide antigens by the MHC molecules on the surface of the antigen presenting cells. The peptides are fragments of proteins or even glycoproteins. The design of the binding groove of the MHC molecule is such that it only binds to peptides in their more or less extended conformation by interaction with anchor residues often towards the terminal ends. Carbohydrates and lipids do not interact with the MHC molecules and they belong to the group of T cell independent antigens unable to elicit a T cell response. However, the elution and analysis of peptides bound to the MHC molecules has revealed that glycan residues remaining from glycoprotein degradation could be attached to peptides bound to the MHC molecules [1]. It is therefore important to investigate the immunogenicity of glycopeptides, in particular the changes observed in the mucin glycosylation pattern in the case of e. g. colon cancer cells due to down regulation of enzymes leading to aberrant glycosylation [2]. Nothing is known about the specificity of the T cell response to glycans attached to peptides binding to MHC. The E^k restricted self peptide, VITAFNEGLK, from CBA/J mouse hemoglobin (67-76) is known to bind well MHC class II molecules, but is unable to stimulate the $CD4^+$ T cells to proliferate [3]. Conversion of such a non-immunogenic peptide into an immunogen by glycosylation would indicate that the T cell specificity was due to direct contact between the glycan and the binding domain of the TCR. Furthermore, by modifying the glycan in this model for T cell-glycan recognition it would be possible to study the detailed molecular interactions between the glycan and the TCR, which in turn could lead to knowledge about the requirements for the efficient design of cancer vaccines.

Results and Discussion

The Hb(67-76) peptide **1**, VITAFNEGLK, was first studied by scanning through the peptide with A, T and S as well as with T(α-D-GalNAc) and T(α-D-GalNAc). These glycopeptides were obtained by multiple column synthesis [4] by use of the conventional Fmoc-amino acid-OPfp ester coupling method with piperidine deprotection on polyethylene glycol polyamide copolymer (PEGA) resin as previously described [5]. As expected any modification of anchor residues abolished both binding to isolated MHC molecules as well as immunogenicity [6]. The V, T, N and L residues, known to point away from the N could be substituted with recognition of the glycan residues. In order to study this

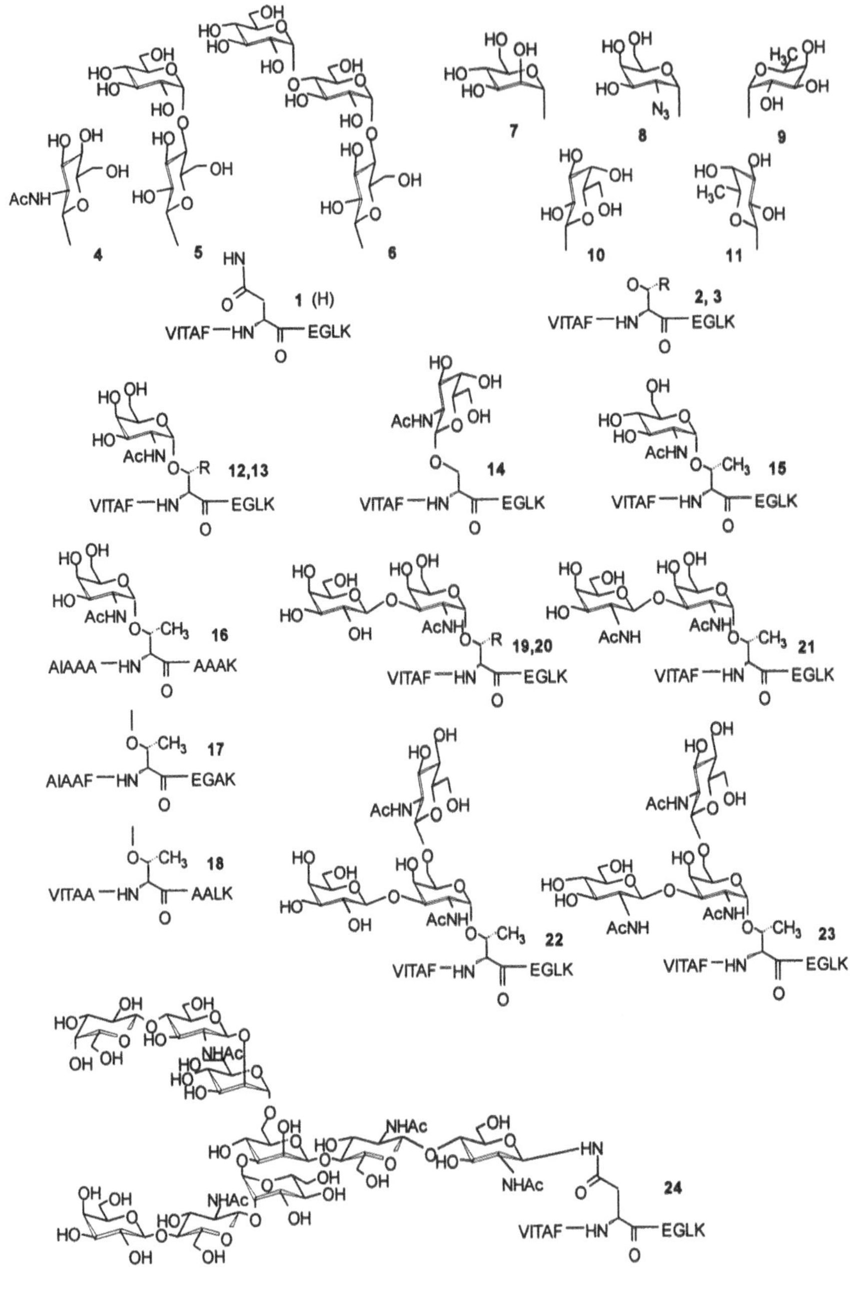

Fig. 1. MHC class II binding N- and O-glycopeptides synthesised and used in T cell proliferation studies. T cell hybridomas were generated against **12** ($R=CH_3$).

T(α-D-GalNAc) and T(α-D-GalNAc) to afford binding to MHC. Furthermore, only the specificity in more glycopeptides with substitution of N^{72} afforded immunogenic peptides MHC molecule could be replaced by an A, S or T, but only the terminal V and the central with specific detail, the large library of glycopeptides presented in Fig. 1 was constructed. The glycopeptides were synthesized by the building block strategy [7]. The smaller glycans were synthesized as previously described.The more complex glycan structures **19-23** were synthesized using *N*-Dts (β-directing) and azido (α-directing) groups as 2-acetamido precursors. The trichloroacetimidates (TCA) were used as glycosyldonors. Ac_3-β-D-GalN_3-TCA was first attached to Fmoc-T/S-OPfp and the product converted to the 4,6 benzylidene derivative. Then glycosylation with Ac_3-β-D-GlcNDts-TCA or Ac_4-β-D-Gal-TCA first in the 3 position of Fmoc-T/S(4,6-*O*-benzylidene-α-D-GalN_3)-OPfp (for **19-21**) and, after removal of benzylidene, in the 6 position gave the branched trisaccharide structures (for **22** and **23**). Two different routes for glycopeptide assembly were compared. These are the previously used method of heterogeneous reduction with Zn in Ac_2O on the building block stage and a method of simultaneous reduction of *N*-Dts group and azide and *N*-acetylation immediately after incorporation of the glycosylated building block on the solid phase. It was found that the new method of azide reduction with DDT on the solid phase was very efficient and the overall yield based on the above trisaccharide precursor was improved. The stepwise reduction and acylation on the solid phase also allowed the incorporation of a photoaffinity label on the amino groups by reaction with 2-azido-Bz-ODhbt. Seven different analogs of GalNAc, GlcNAc and core1 were photoaffinity labeled and found to be immunogenic. The large branched structure **24** was obtained from fetuin by hydrazinolysis to yield its glycan pool. After conversion to the glycosylamines these were reacted with Fmoc-Asp(ODhbt)-OtBu in DMSO and building blocks obtained in high yield were separated by HPLC. By conversion into peracetates and TFA cleavage of the tBu ester the protected building blocks could be obtained. Incorporation into VITAFNEGLK afforded the MHC binding glycopeptide.

The binding of the glycopeptides to MHC and the ability to elicit T-cell proliferation were

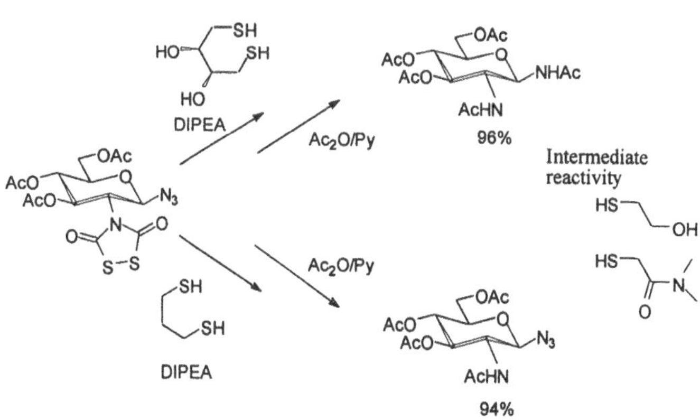

Fig. 2. The reduction of the N-Dts protecting group can be performed selectively or simultaneously with an azide group.

investigated. It was found that all the glycopeptides **4-23** bound to MHC E^k with an affinity comparable to **1** while the ability to stimulate the immune system varied greatly and the stimulation did not depend on MHC binding affinity. Particularly immunogenic were the glycopeptides **10, 12, 13, 14,** and **20** (R=H). Very analogous glycopeptides **7, 8, 15, 20** ($R=CH_3$) and the core2 structure **22** where a GlcNAc is introduced on the 6-position of GalNAc were essentially non-immunogenic, indicating close molecular contact between glycan and the TCR and a particular potency of the natural aberrant glycosylations to stimulate T cell proliferation. 120 hybridoma T cell lines were produced against **12** ($R=CH_3$) and the selectivity of their T cell receptor towards the glycan and peptide part was investigated. The hybridomas recognized both parts of the glycan and parts of the peptide. The frequency of cross-reactivity towards S(α-GalNAc) **13** was quite high but some clones showed selectivity to **12.** In most cases even minor change in the glycan eliminated cross-reactivity. Three out of the 19 selected glycopeptide specific hybridomas cross-reacted with core1 structure **19.** When the 4-hydroxyl epimer **15** was tested only 2 hybridomas were reactive. With introduction of GlcNAc as in the mucin core2 structure **22** abolished the immunogenicity and no cross-reactivity was found. This result could be significant for the design of glycopeptide vaccines against cancer.

Acknowledgments

This work was supported by The Danish Cancer Association and The EC Science Program (Grant MM SCI*-CT92-0765).

References

1. Chicz, R. M., Lane, W. S., Robinson, R. A., Trucco, M., Strominger, J. L. and Gorga, J. C., Intern. Immunol., 6 (1994) 1639.
2. Meldal, M., Holm, C. B., Bojesen, G., Jacobsen, M. H. and Holm, A., Int. J. Pept. Prot Res., 4(1993) 250.
3. Evavold, B. D., Williams, S. G., Hsu, B. L., Buus, S. and Allen, P. M., J. Immun., 148 (1992) 347.
4. Yang, J.-M., Byrd, J. C., Siddiki, B. B., Chung, Y.-S., Okuno, M., Sowa, M., Kim, Y. S., Matta, K. L. and Brockhausen, I., Glycobiology, 4 (1995) 873.
5. Frische, K., Meldal, M., Werdelin, O., Mouritsen, S., Jensen, T., Galli-Stampino, L. and Bock, K., J. Pept. Sci., 1996 in press.
6. Jensen, T., Galli-Stampino, L., Mouritsen, S., Frische, K., Peters, S., Meldal, M. and Werdelin, O., Eur. J. Immunol., 26 (1996) 1342.
7. Meldal, M. and Bock, K., Glycoconjugate J., 11 (1994) 59.

Fluorescent quenched libraries for monitoring chemical and enzymatic disulfide bond formation and cleavage

J.C. Spetzler[a], V. Westphal[b], J.R. Winther[b] and M. Meldal[a]
[a]Department of Chemistry, [b]Department of Yeast Genetics, Carlsberg Laboratory, Gamle Carlsberg Vej 10, DK-2500 Valby, Copenhagen, Denmark

Introduction

In the past 10 years, peptide libraries have been used extensively for drug discovery [1-3]. However, a less well known aspect is their application for the study of chemical and enzymatic reactions. In the present work, a peptide library has been developed and the cleavage and formation of disulfide bridges using enzymes and chemical reagents was investigated. The basic design of the library consists of two chromophores, a fluorescent probe, the o-aminobenzoyl group (Abz) and a quencher, the Tyr(NO_2) moiety, which allow sensitive monitoring of both disulfide bond formation and cleavage in solution and on solid phase, respectively [4]. The two chromophores are incorporated on separate peptide charts which are connected through an inter-chain disulfide bridge. We here describe the peptide library as well as two disulfide bond containing model systems which have been applied as substrates in chemical and enzymatic S-S reduction and demonstrate the feasibility of developing such libraries.

Results and Discussion

A major prerequisite for the construction of a peptide library which contains defined disulfide bonds as well as a built-in monitoring system is that its chemistry is well established. Therefore, two inter-chain disulfide bond containing model systems were synthesized to be used in both solution and on solid phase using PEGA-beads [5] (**1** and **2** in Fig. 1). In the model systems on solid phase, the peptide was linked via Met so that the peptides can be selectively cleaved from the beads using CNBr. In the first model system (Fig.1), an Abz group was incorporated at the N-terminal of one peptide and another peptide contained a Tyr(NO_2) residue and an N-terminal Cys residue. The inter-chain disulfide bond was formed by the reaction between the 3-nitro-2-pyridine sulfenyl (Npys) derivative of Cys and the thiol moiety of Cys [7]. For testing whether the nature of the amino terminal would influence the S-S formation and cleavage, a second model system was developed (**2** in Fig. 1). This consisted of a peptide with an Abz group at the N-terminal. This peptide was connected via an inter-chain disulfide bridge to a second peptide which also contained the Tyr(NO_2) residue and an internal Cys residue. The inter-chain disulfide bridge was formed from the thiol moiety of Cys and the 5-nitro-2-pyridine sulfenyl (*p*Npys) derivative of Cys [8].

Both model systems were used as substrates in the investigation of chemical S-S reducing reagents such as thiols, phosphines and silanes in solution and on beads, respectively. Since these reagents could have been able to reduce the Tyr(NO_2) moiety, the stability of model system 1 in the presence of reducing reagents was first investigated using ^{1}H-NMR.

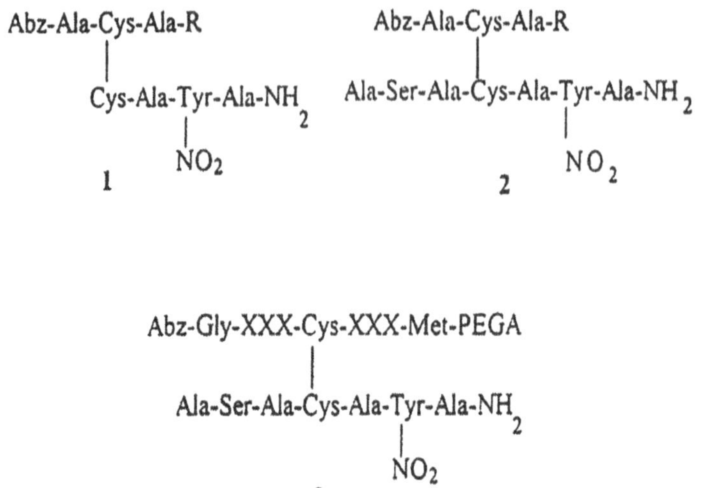

3

R=NH_2 or Met-PEGA resin
X=all the natural amino acids except
Abz=o-amino benzoyl

*Fig. 1. Structure of model systems **1** and **2** (top) and the peptide library **3** (bottom).*

Fortunately, these studies show no evidence of reduction of the quenching chromophore [6].

The S-S reduction in solution was monitored by use of a fluorescence spectrophotometer, whereas on the beads, the reduction was visually monitored as the colour on the beads changed from dark to bright blue. In these experiments, a 50 and a 10 fold excess of reagents were used for monitoring the S-S reduction rate under optimal pH conditions in solution and on the beads, respectively. In general, model system **2** was reduced comparatively faster than model system **1** both in solution and on beads (Fig. 2 and 3). The order of the efficiency of the reagents was dithiothreitol (DTT) > 2-mercaptoethanol (ME)>> tris(2-carboxyethyl) phosphine (TCEP) whereas triisopropyl silane (TIS) was not able to reduce the S-S bond.

A peptide library (Fig. 1, **3**) was synthesized by split synthesis in order to yield one compound on each bead and consisted of an internal Cys residue which was flanked by one randomized amino acid on one side and three randomized amino acids on the other side (all the natural amino acids except Cys in these positions). The Abz group was incorporated at the N-terminal of the peptide. The library was linked to PEGA-beads via Met.

A disulfide bridge was formed between the bead linked library and a peptide containing Tyr(NO_2) and Cys(*p*Npys) residues. The formation and cleavage of the disulfide bridges in the peptide libraries were monitored by viewing the color change of the beads under a fluorescence microscope. Beads which deviate from the average could be collected and analyzed by amino acid sequencing.

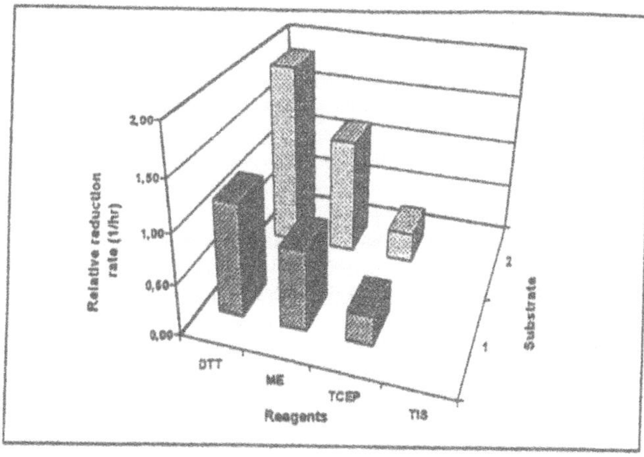

Fig. 2. Monitoring of the S-S reduction in solution using a 50 fold excess of the reducing reagent in 1mM EDTA, pH 7.5 (for DTT, ME and TIS) and 1mM NaOAc, pH 4.5 (for TCEP). The concentration of each substrate was 1 μM

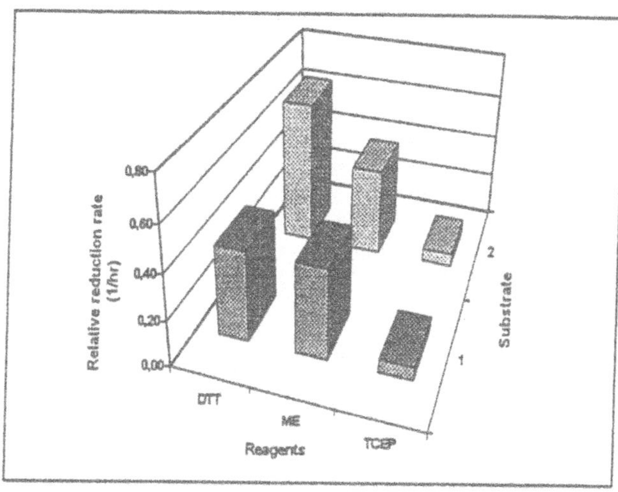

Fig. 3. Monitoring of the S-S reduction on beads using a 10 fold excess of the reducing reagent in 1mM EDTA, pH 7.5 (for DTT and ME) and 1mM NaOAc, pH4.5 (for TCEP).

The formation of correct disulfide bonds is essential for the folding of many proteins. Protein disulfide isomerase (PDI) catalyzes the oxidative folding of unfolded proteins and acts by catalyzing thiol-disulfide exchange reactions which can lead to formation, reduction or isomerization of protein disulfide bonds [9]. Due to the complex nature of protein folding reaction reliable assays for PDI have suffered from low sensitivity and poor reproducibility. An assay allowing direct observation of peptide oxidation and determination of disulfide bond forming activity using these model systems is currently under investigation.

In conclusion, we have developed a peptide library which can be a powerful tool in the study and improvement of enzymatic and chemical disulfide bond reactions.

References

1. Lam, K.S., Salmon, S.E., Hersh, E.M., Hruby, V.J., Kazmierski, W.M. and Knapp, R.J., Nature, 354 (1991) 82.
2. Houghten, R.A., Pinilla, C., Blondelles, S.E., Appel, J.R., Dooley, C.T. and Cuervo, J.H., Nature, 354 (1991) 84.
3. Furka, A., Sebestyen, F., Asgedom, M. and Dibo, G., Int. J. Pept. Prot. Res., 37 (1991) 487.
4. Meldal, M. and Breddam, K., Analytical Biochemistry, 195 (1991) 141.
5. Meldal, M., Tetrahedron Lett., 33 (1992) 3077.
6. Spetzler, J.C., Westphal, V., Winther J.R .and Meldal, M. (in preparation).
7. Albericio, F., Andreu, D., Giralt, E., Navalpotro, C., Pedroso, E., Ponsati, B and Ryiz-Gayo, M., Int. J. Pept. Prot. Res., 34 (1989) 124.
8. Rabanal, F., DeGrado, W.F. and Dutton, L., Tetrahedron Lett., 37 (1996) 1347.
9. Freedman, R.B., Hirst, T.R. and Tuite, M.F., TIBS, 19 (1994) 33.

Simultaneous multiple synthesis by the Multipin™ method: Techniques for multiple handling, high throughput characterization and reaction optimization on solid phase

Andrew M. Bray, N. Joe Maeji, Liana M. Lagniton, Debra Chiefari, Ian W. James, Kiah Ang, Nicolas Ede, Robert M. Valerio and Tom J. Mason
Chiron Mimotopes Pty. Ltd., 11 Duerdin Street, Clayton, Victoria 3168, Australia

Introduction

The Multipin™ method of multiple solid phase synthesis allows access to large numbers of individual compounds (100s - 1000s, i.e. libraries [1]) via parallel synthesis, at scales ranging from 1 to 40 μmol/pin. Synthesis is performed on polymer grafted plastic pins, which are organized in an 8 x 12 modular format. The method was first developed for the rapid elucidation of linear antibody epitopes [2], via the screening of sub-micromolar quantities of pin-bound peptide segments. With the development of new graft polymers for pins, such as poly(hydroxyethyl methacrylate) [3], acrylamide copolymers [4], and polystyrene [4], the development of high loading pin designs [5], and the introduction of a wide range of linker systems, the Multipin™ method can now be applied to a far wider range of synthetic problems. Since Ellman's landmark paper describing the synthesis of benzodiazepines on resin [6], there has been a dramatic growth in the field of multiple solid phase organic synthesis [1,7,8] and several examples of non-peptide synthesis on pins have been published [9-12]. Unlike solid phase peptide or oligonucleoside synthesis, solid phase organic synthesis requires that a very broad range of reaction types must proceed in high yield. Although an 80% conversion per step is adequate in solution phase synthesis (where intermediates can be purified) solid phase synthesis requires that far higher conversions be obtained in order to achieve adequate target compound purities. Apart from enabling parallel synthesis of target compounds, multiple solid phase synthesis can be used as a tool to explore reaction conditions [13]. We therefore use multiple synthesis on pins to screen reaction conditions in order to find optimum chemistries rapidly [11], which is achieved by synthesizing many copies of a given compound, while varying all critical reaction parameters. This coupled with the ability for multiple cleavage [14], high throughput characterization by ion spray MS [15] and HPLC, and computer assisted data assessment enables rapid comparison of reaction conditions. Using this powerful methodological strategy, our goal is to develop simple yet highly efficient synthetic methods for use in solid phase synthesis and in particular, in the synthesis of large sets of compounds for drug lead discovery.

Results and Discussion

Efficient simultaneous multiple synthesis requires that all handling steps be performed either in parallel, or in a high-throughput serial manner. At early stages of synthesis, where a common handle and possibly template are incorporated onto the pin surface, it is most convenient to perform chemistry on the heads of pins (crowns) prior to their attachment to the 96 position pin holder, using wide-neck standard laboratory glass-ware. Once crowns are assembled onto the 96 position pin holder, many multiple handling procedures are easily performed. Common reaction steps and washing steps can be performed in polypropylene baths. Parallel reactions requiring different reagents can be performed in polypropylene 96 well microtitre trays. The removal of reagent or washing solvent from the pins is achieved simply by shaking the pin assembly, followed by additional washes if required. Hence the tedious task of multiple filtration is avoided. Although this equipment is adequate for most chemistries associated with peptide assembly, many reactions of potential interest for the solid phase synthesis of nonpeptide targets must be performed under an inert atmosphere, and/or require refluxing conditions, etc. The challenge then is to develop simple handling techniques for multiple synthesis. For example, setting up 96 refluxing reactions is a difficult task. Rather than design specialized equipment for this process, a reaction can be optimized in a very high boiling point solvent at a temperature well below the boiling point. Now without the need to contain a refluxing solvent, large numbers of reactions can be performed at elevated temperature using pins placed into racked screw capped vials, and a standard laboratory oven. This is well illustrated by a solid phase Suzuki cross coupling [16,17] study where phenylboronic acid was coupled to the peptide model system p-Br-C_6H_4-CO-Ala-Phe-(Rink Handle [18])-Pin (loading = 1.5 μmol/pin) in the presence of $Pd(Ph_3P)_4$ and Na_2CO_3. The screening of a number of high boiling point solvents indicated that ethoxyethanol was an excellent solvent for this reaction. When the reaction was performed in an oven at 63°C for 18 h, using ethoxyethanol (b.p. = 135°C) as solvent, p-Ph-C_6H_4-Ala-Phe-NH_2 was obtained in >90% purity by HPLC (peak area at 214 nm) upon cleavage. This method has been subsequently used in biaryl library synthesis. Suzuki chemistry is sensitive to oxygen. Once again, rather than design specialized apparatus for performing multiple air sensitive reactions, relatively simple equipment is used. All handling prior to sealing the racked vials, including multiple dispensing with an automated solvent resistant pipettor is performed under argon in an atmosphere bag (e.g. Aldrich Z11,836-2). The general strategy of simple handling techniques allows for routine multiple syntheses by individual chemists.

The assembly of target compounds is followed by multiple cleavage. Cleavage and post cleavage handling must be streamlined to allow for the handling of large numbers of compounds. We have developed a number of cleavage strategies that are useful tools in multiple peptide synthesis: cleavage by diketopiperazine-forming handle at neutral pH [19], gas phase ammonolysis followed by multiple elution [20] and multiple cleavage of esters with NaOH in MeCN(aq) [21]. Multiple cleavage by acidolysis [14] however, is still currently the most useful cleavage strategy for use in small molecule synthesis. In general, Multipin™ cleavage is performed in racked polypropylene tubes. The efficiency of the first three methods listed above is greatly enhanced if cleavage (or elution) is performed in an ultrasonic field [22]. Sonication coupled with TFA-based cleavage is too harsh for the pin graft polymers. Cleavage is started simultaneously by placing 96 pins into racked tubes containing cleavage reagent. Work up (i.e. removing the solid phase from the cleavage solution) is performed by simply lifting the block of 96 mounted pins from the racked tubes into which cleavage was performed. A vacuum centrifuge such as a Savant SC210A is capable of evaporating four sets of 96 cleavage solutions (384 solutions) simultaneously.

Removal of residual TFA used in acidolysis is achieved by resuspending cleavage products in MeCN(aq) and then evaporating the resulting solutions. Products prepared with this post-cleavage handling are not inherently toxic in cellular assays [23].

After target compounds are cleaved from the pin surface, characterization is necessary. Since most characterization methods handle one sample at a time, high throughput methods are required. Techniques such as automated HPLC and ion spray MS are very useful in this regard. Interestingly, we have found good agreement between peptide purities determined by HPLC and ion spray MS [15]. There is also a growing interest in characterization of support-bound products. For example, FTIR has been applied to resin samples with good results [24]. FTIR has also been successfully applied to pins [25]; in this case, the pin is not damaged, so it can be analyzed repeatedly throughout the course of a synthesis. This method promises to be an effective high throughput analytical tool.

Combining simultaneous multiple synthesis by the Multipin™ method with multiple cleavage and high throughput characterization, gives rise to a powerful methodological strategy for developing high efficiency reactions for use in solid phase synthesis. When searching for a general method, it is useful to select model systems that are easily assembled, that are easily quantitated, and that yield targets that are also easily quantitated. We have found simple peptide models of the form X-Ala-Phe-(Linker)-Pin (where X contains the functional group(s) undergoing transformation) to be easily analyzed by HPLC and ion spray MS [11]. In our study on reductive amination of support-bound ketones [11], a model system containing 4-oxoproline was used. Five parameters were simultaneously varied in the optimization study: amine, amine concentration, solvent, pH, and $NaBH_3CN$ concentration. A total of 56 conditions were explored concurrently. Following multiple cleavage and high throughput analysis, the % conversion to target was calculated for each condition. Optimum conditions were then identified by rank ordering the data in decreasing order of % conversion. The optimized conditions were demonstrated to be highly efficient for a wide range of amines, including poor nucleophiles such as p-bromoaniline.

This general approach is also useful in linker development. For example, in developing methods to attach phenols directly onto pin-bound hydroxymethylphenoxyacetyl handle [26] via Mitsunobu chemistry, the model system *p*-$(HOCH_2)C_6H_4$-CO-Ala-Phe-(Rink Handle)-Pin was used. As the ether bond formed upon phenol coupling is stable to acidolysis, coupling efficiency can be determined by comparing the ratio of cleaved product [*p*-$(Ar\text{-}OCH_2)C_6H_4$-CO-Ala-Phe-NH_2] with cleaved starting material [*p*-$(HOCH_2)C_6H_4$-CO-Ala-Phe-NH_2]. Once optimum conditions were identified, the method could be applied to the handle of interest.

In conclusion, the Multipin™ strategy is based on simple handling techniques which greatly facilitate multiple handling. The development of new multiple handling techniques now allows the method to be applied to solid phase organic synthesis. The combination of multiple synthesis and high throughput characterization allows for the rapid screening of reaction conditions, hence providing researchers with a powerful tool for developing and adapting chemistry for use in solid phase organic synthesis.

References

1. Gordon, E.M., Barrett, R.W., Dower, W.J., Fodor, S.P.A. and Gallop, M.A., J. Med. Chem., 37 (1994) 1385.

2. Geysen, H.M., Barteling, S.J. and Meloen, R.H., Proc. Natl. Acad. Sci. USA, 82 (1985) 178.
3. Valerio, R.M., Bray, A.M., Campbell, R.A., DiPasquale, A., Rodda, S.J., Geysen, H.M. and Maeji, N.J., Int. J. Peptide Protein Res., 42 (1993) 1.
4. Maeji, N.J., Valerio, R.M., Bray, A.M., Campbell, R.A. and Geysen, H.M., Reactive Polymers, 22 (1994) 203.
5. Maeji, N.J., Bray, A.M., Valerio, R.M. and Wang, W., Peptide Research, 8 (1995) 33.
6. Bunin, B.A. and Ellman, J.A., J. Am. Chem. Soc., 114 (1992) 10997.
7. Terrett, N.K., Gardner, M., Gordon, D.W., Kobylecki, R.J. and Steele, J., Tetrahedron, 51 (1995) 8135.
8. Thompson, L.A. and Ellman, J.A., Chem. Rev., 96 (1996) 555.
9. Bunin, B.A., Plunkett, M.J. and Ellman, J.A., Proc. Natl. Acad. Sci. USA, 91 (1994) 4708.
10. Virgilio, A.A. and Ellman, J.A., J. Am. Chem. Soc., 114 (1994) 11580.
11. Bray, A.M., Cheifari, D.S., Valerio, R.M. and Maeji, N.J., Tetrahedron Lett., 36 (1995) 5081.
12. Valerio, R.M., Bray, A.M. and Patsiouras, H., Tetrahedron Lett., 37 (1996) 3019.
13. Hudson, D., J. Org. Chem., 53 (1988) 617.
14. Valerio, R.M., Bray, A.M. and Maeji, N.J., Int. J. Peptide Protein Res., 44 (1994) 158.
15. Smart, S.S., Mason, T.J., Bennell, P.S., Maeji, N.J. and Geysen, H.M., Int. J. Peptide Protein Res., 7 (1996) 47.
16. Backes, B.J. and Ellman. J.A., J. Am. Chem. Soc., 116 (1994) 11171.
17. Frenette, R. and Friesen, R.W., Tetrahedron Lett., 35 (1994) 9177.
18. Rink, H., Tetrahedron Lett., 28 (1987) 3787.
19. Bray, A.M., Maeji, N.J., Valerio, R.M., Campbell, R.A. and Geysen, H.M., J.Org. Chem., 56 (1991) 6659.
20. Bray, A.M., Jhingran, A.G., Valerio, R.M. and Maeji, N.J., J. Org. Chem., 59 (1994) 2197.
21. Valerio, R.M., Benstead, M., Bray, A.M., Campbell, R.A. and Maeji, N.J., Anal. Biochem., 197 (1991) 168.
22. Bray, A.M., Lagniton, L.M., Valerio, R.M. and Maeji, N.J., Tetrahedron Lett., 35 (1994) 9079.
23. Mutch, D.A., Underwood, J.R., Geysen, H.M. and Rodda, S.J., J. AIDS, 7 (1994) 879.
24. Yan, B. and Kumaravel, G., Tetrahedron, 52 (1996) 843.
25. Gremlich, H.-U. and Berets, S.L., App. Spectroscopy, 50 (1996) 532.
26. Atherton, E., Logan, C.L. and Sheppard, R.C., J. Chem. Soc. Perkin Trans. I, (1981) 538.

Synthesis and antitumour activities of analogs and segments of *Papaver somniferum* pollen tridecapeptide

Jia-Xi Xu and Sheng Jin
Department of Chemistry, Peking University, Beijing 100871, China

Introduction

Finding novel active compounds from plants is still a very important path in the search for new pharmaceutical materials which can be used as lead compounds for developing new drugs. Pollen possesses many physiological activities and plays an important role during plant reproduction [1]. It therefore should contain key substances, including small peptides. In recent years, our group has studied several kinds of pollen and obtained several peptides. The physiological activity studies showed that they had some characteristics promoting immunity and restraining cancer cell division [2-5].

The tridecapeptide NQNGSNPKTVKQA, named PSPP3, was obtained from *Papaver somniferum* pollen. Biological assay results indicated that it showed some anti-tumour activity [6]. In this paper, we report the synthesis and anti-tumour activity of analogs and segments of this tridecapeptide against human leukaemia tumour cell HL-60 and human red leukaemia tumour cell K562.

Results and Discussion

In recent investigation, we have found that the active peptides separated from *Brassica campestris* and *Papaver somniferum* pollen have some sequence similarities [2-4]. They have many amino acid residues containing carboxamide groups on side-chain and some consecutive ones in their N- or C-terminals. In order to find effective anti-tumour agents and the relationship of structure and activity, firstly five analogs were designed by replacing one or more N or Q residues with D or E, respectively to find out whether carboxamide groups on side-chains are important to anti-tumour activities. Secondly, decapeptide segment PSPP3-4 deducted three consecutive amino acid residues NQN from N-terminal and its analog PSPP3-8, in which N and Q were replaced with D and E, respectively, were designed to make clear whether these three N-terminal amino acid residues are important to anti-tumour activities. Heptapeptides based on N- and C-terminals and analog derived from N-terminal heptapeptide were also designed for finding the active site. Their sequences are given as follows.

PSPP3-1, DEDGSNPKTVKQA
PSPP3-2, DENGSNPKTVKQA
PSPP3-3, NENGSNPKTVKQA
PSPP3-4, GSNPKTVKQA
PSPP3-5, PKTVKQA
PSPP3-6, DEDGSDPKTVKEA
PSPP3-7, NQNGSDPKTVKEA

PSPP3-8, GSDPKTVKEA
PSPP3-9, NQNGSNP
PSPP3-10, DEDGSNP

These designed peptides were synthesized using the Merrifield solid phase peptide synthesis method with the acid-labile tert-butyloxycarbonyl group for temporary protection, acid-stable groups for side chain protection, and DCC/HOBt or HOSu as coupling reagent [2,5]. Peptides were cleaved by anhydrous hydrogen fluoride with anisole and thioanisole as scavengers and washed with cooled ethyl ether. Peptides were extracted with 30% acetic acid and purified by gel filtration on Sephadex G10 and G15 columns, and further purified by reverse-phase HPLC if necessary. The purity of the peptides was checked by analytical HPLC and the composition was confirmed by amino acid analysis.

Table 1. Antitumour Activities of Synthetic peptides

Peptide	Concentration (M)	Inhibitive Ratio to HL-60(%)	Inhibitive Ratio to K562(%)
PSPP	10^{-4}	24.8	-5.78
	10^{-5}	25.2	16.0
	10^{-6}	42.5	18.7
	10^{-7}	50.4	39.0
PSPP3-1	10^{-4}	-5.9	-34.9
	10^{-5}	14.1	6.69
	10^{-6}	52.4	26.6
	10^{-7}	52.7	27.7
PSPP3-2	10^{-4}	12.1	-26.7
	10^{-5}	25.5	-4.78
	10^{-6}	35.3	9.96
	10^{-7}	35.3	15.9
PSPP3-3	10^{-4}	0.60	-7.17
	10^{-5}	1.10	-14.7
	10^{-6}	7.90	9.48
	10^{-7}	12.6	20.3
PSPP3-4	10^{-4}	-1.80	-31.3
	10^{-5}	3.45	-46.7
	10^{-6}	12.5	11.8
	10^{-7}	18.2	13.3
PSPP3-5	10^{-4}	18.3	-62.9
	10^{-5}	19.7	-25.0
	10^{-6}	28.1	1.27
	10^{-7}	28.5	4.06
PSPP3-6	10^{-4}	12.6	-70.0
	10^{-5}	16.5	46.1
	10^{-6}	31.4	41.2
	10^{-7}	44.8	48.3

Table 1. (continued)

Peptide	Concentration (M)	Inhibitive Ratio to HL-60(%)	Inhibitive Ratio to K562(%)
PSPP3-7	10^{-4}	3.20	-34.1
	10^{-5}	10.9	2.07
	10^{-6}	27.7	41.3
	10^{-7}	35.3	58.9
PSPP3-8	10^{-4}	21.9	-20.1
	10^{-5}	26.0	39.4
	10^{-6}	40.6	33.0
	10^{-7}	45.9	37.6
PSPP3-9	10^{-4}	7.51	-2.05
	10^{-5}	25.8	31.7
	10^{-6}	25.4	33.0
	10^{-7}	50.2	37.8
PSPP3-10	10^{-4}	-2.86	5.78
	10^{-5}	1.45	11.2
	10^{-6}	20.0	-0.60
	10^{-7}	27.2	19.4

Anti-tumour activities of all synthetic peptides above were assayed by determining the inhibitive concentrations of peptides to human leukaemia tumour cell HL-60 strain with the tetrazolium salt MTT(3-(4,5-dimethylthiazol-2-yl)-2,5-diphenyl tetrazolium bromide) method [7]. Bioassay results indicated that analogs PSPP3-1 and N-terminal segment heptapeptidePSPP3-9 showed almost the same antitumour activities as the natural tridecapeptide. However, analogs PSPP3-2, PSPP3-6, PSPP3-7 and segment PSPP3-8 showed slightly lower activities than natural peptide. Other synthetic peptides showed very low activities. The conformations of all analogs, segments and PSPP3 were determined by CD spectroscopy. According to CD spectra, PSPP3-7 showed mainly β-turn conformation. The others showed β-sheet conformation. However, PSPP3-1, PSPP3-3 and PSPP3-4 also showed partial β-turn conformation. It seems that the N-terminal segment is important and β-turn conformation is not favorable to anti-tumour activities against human leukaemia tumour cell HL-60.

Replacing the amino acid residues Q and N with E and D at the C-terminal and middle segment led to increased inhibition of the proliferation of human red leukaemia tumour cell K562 strain. N-terminal heptapeptide PSPP3-9 and modified C-terminal decapeptide PSPP3-8, that had N to D and Q to E replacements at C-terminal part, showed almost the same activities as natural tridecapeptide. Similar modifications at N-terminal part of the peptide led to reduced activity, C-terminal heptapeptide PSPP3-5 showing the lowest activity. β-turn conformation seems to favour anti-tumour activities against human red leukaemia tumor cell K562.

Acknowledgments

The project was supported by National Natural Science Foundation of China and Youth

Foundation of Peking University.

References

1. Wang, K. F., The nutritive value and food therapy of pollen, Peking University Press, Beijing, 1986.
2. Xu, J. X., Jin, S., Chin. Chem. Letts., 4 (1993) 213.
3. Zhang, Y. L. and Jin, S., *ibid.*, 1 (1990) 85.
4. Jin, S., Xu, J. X. and Miao, P., Youji Huaxue, 13 (1993) 202.
5. Xu, j. X. andJin, S., Acta Chimica Sinica, 53 (1995) 822.
6. Xu, J. X., Jin, S. andCai, M. S., Chin. Chem. Letts., 6 (1995) 941.
7. Denizot, R. andLang, R., J. Immunol. Methods., 89 (1986) 271.

Utilization of phage displayed peptides in the purification of genetic engineering products

Rui Fang, Jie Qi, Tian-Yu Li, Hui Zhou and Wei Li
Department of Molecular Biology, Jilin University, Changchun 130023, China

Introduction

Peptide library techniques have been developed rapidly and applied widely in various fields of molecular recognition, antigen determination and vaccine and drug design [1-3]. In these studies, phage displayed ligands or synthetic peptide libraries were used to select functional ligands for target molecules, such as antibodies, enzymes and cellular receptors etc. In this paper, we explored the potential use of a peptide library in preparing novel affinity-chromatographic materials to improve the recovery of genetic engineering products. The main idea is to utilize the peptide library to select high affinity ligands to the target protein, α-INF in this work, and to prepare an affinity-chromatographic matrix with immobilized phage displayed peptide ligands.

Results and Discussion

Ligands of α-INF were selected by four rounds of biopanning (methods as described by G.P. Smith [4]) using a phage hexa-peptide library. The recoveries of phages (Table 1) indicate that phage peptides with comparatively high affinity to α-INF have been condensed, since the recovery of phages increased while the concentration of biotin-INF was decreased. The low recovery of the fourth round may be due to too low INF concentration and too high amount of input phages.

Table 1. Recoveries of four rounds of biopanning of α-INF

	input of biotin-INF (μg)	input of phages (TU)	recovery (%)
first round	60	$1.0\text{x}\ 10^{9}$	$2.5\text{x}10^{-4}$
second round	6	$5.0\ \text{x}10^{10}$	$1.0\text{x}10^{-2}$
third round	1.2	$4.0\text{x}10^{10}$	$6.2\text{x}10^{-2}$
fourth round	0.12	$6.0\text{x}10^{11}$	$1.2\text{x}10^{-4}$

First round by method (P+LS), others by (PL+S) [4].
TU: tetracyline transducing unit of recombinant phages

Table 2. Comparison of phages selected and wild type phage fl

	input phage	output phage	recovery (%)
phages selected out	6.0×10^7	2.2×10^5	0.37
fl	1.0×10^8	250	4.0×10^{-4}

To demonstrate the specific affinity of phage peptide to α-INF, we compared the of recombinant phages and wild type phage fl in micropanning. The result shows that the recovery of phages that displayed peptide ligands of α-INF is about 1000 times higher than that of fl as a result of non-specific interactions.

We analyzed INF-affinity of both the phages selected and the antibody. The result (Fig.1) shows that 0.4×10^{12} TU/ml phage and 1.25 μg/ml antibody gave signals of similar intensity. As each phage particle carries five copies of peptide ligands fused to pIII, we can roughly estimate that the affinity of phages is about 20 times lower than that of antibody, but could probably be improved by further screening for better ones in the phage mixture.

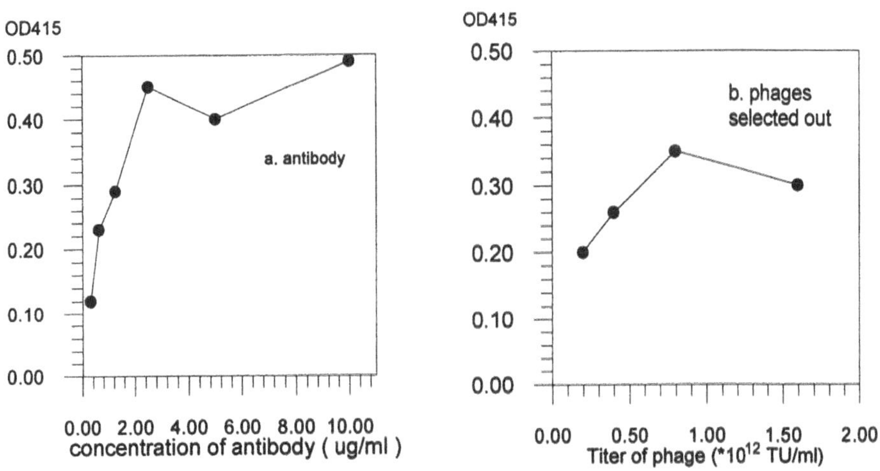

Fig. 1. Comparison of INF-affinity of phage peptides with anti-INF antibody. 30 μl diluted phage and antibody in PBS were coated on a 40-well plate by drying. After the plate was blocked by 1%BSA, biotin-INF (1 μg/ml, in PBS/Tween) was added at 43°C for 30 min. INF bound to the plate was detected by Avidin-HRP. a. result of anti-INF antibody b. result of phages selected.

Acknowledgment

α-INF and anti-INF antibody were kindly provided by Prof. Liu Suying of Changchun Biology Product Institute.

References

1. Felici, F., Castagnoli, L., Musacchio, A. and Jappelli, R., J. Mol. Biol., 222 (1991) 301.
2. Eichler J. and Houghten R. A., Biochemistry, 32 (1993) 11035.
3. Fang R., Qi J., Lu Zhibin, Li W.and Shen J.C., Biochem. Biophys. Res. Comm., 220 (1996) 53.
4. Scott, J. K. and Smith, G. P., Science, 249 (1990) 386.

A biologically active peptide isolated from the buckwheat pollen

Yan-Ling Song[a], Xia Wang[a], Shu-Wen He[a], Sheng Jin[a], Jun-Da Liu[b] and Shu Wang[b]

[a] *Department of Chemistry, Peking University, Beijing 100871, China*

[b] *Department of Bio-message and Modulation, Beijing Hospital, Institute of Geriatrics, Beijing 100730, China*

Introduction

Pollen is one type of male reproductive organ in higher plants. During our systematic study on active peptides isolated from serial Chinese pollen, we found that pollen peptides have many physiological activities, for example, promoting immunity and restraining cancer cell division. At CPS-94, we reported the synthesis and the characterization of Chinese rape pollen peptide RPP [1]. Recently a new basic peptide BPP-1 has been isolated and identified from the Chinese buckwheat pollen. The peptide BPP-1 and its two analogues have been synthesized by Merrifield SPPS method [2].

Because the content of certain active peptides in the pollen is very low, so the purification of pollen peptides is difficult and tedious. In order to obtain a little pure peptide, we have to deal with a great quantity of pollen. In this report, we suggest a rapid and efficient method to isolate and purify pollen peptides, and have studied the effect of these peptides on spleen lymphocytes.

Results and Discussion

For obtaining the active peptide BPP-1 from Chinese buckwheat pollen, various separation methods were studied carefully. The extraction solution of two kilograms of buckwheat pollen was processed through ion-exchange chromatography to remove saccharides after removing the proteins contained in the pollen. That is a rapid and efficient method for the isolation of peptides from pollen. The isolated fraction is purified on Dowex 1x2-100 strongly basic anion exchange chromatography to remove pigments and non-peptide substances. Then the isolated fraction is separated into six fractions containing peptide on Dowex 50x2-200 strongly acid cation exchange chromatography. The sixth fraction was purified on RP-HPLC using a linear gradient elution. We obtained the peptide BPP-1 (1.4mg) whose purity was confirmed by dansyl chloride *N*-terminal determination and amino acid analysis. The results showed that its *N*-terminal amino acid residue is alanine and has areasonable ratio of amino acid (shown in Table 1). The molecular weight of BPP-1 is confirmed by FABMS (shown in Table 1). The primary analysis of the peptide BPP-1 using 4-*N,N*-dimethylaminoazobenzene-4' isothiocyanate / phenyl-isothiocyanate double coupling method [3] showed the sequence of a dodecapeptide (listed in Table 2). The peptid BPP-1 and its analogues were synthesized by Merrifield SPPS method with *tert*-butyloxycarbonyl (Boc) group for temporary amino-terminal protection.The following side-chain protection groups were used: O-benzyl for Thr, Ser; N^{ε} -benzyloxycarbonyl for Lys.

Peptide resin was cleaved by hydrogen fluoride and extracted by 10% acetic acid in water. The peptides were purified by RP-HPLC using a linear gradient elution. The results of AAA and FABMS are represented in Table 1.

Table 1. Amino acid analysis and FABMS of peptides BPP

	Ala	Pro	Val	Leu	Gln	Ile	Lys	Thr	Gly	Ser	Asn	FABMS Found	FABMS Calculate
BPP-1 Synthesized	0.93 (1)	0.94 (1)	0.82 (1)	0.96 (1)	1.05 (1)	0.92 (1)	1.98 (2)	0.91 (1)	0.97 (1)	0.72 (1)	0.80 (1)	1257	1255.52
BPP-1 Extracted	1.02 (1)	1.07 (1)	1.10 (1)	1.05 (1)	1.16 (1)	0.95 (1)	1.99 (2)	1.01 (1)	1.05 (1)	0.92 (1)	1.06 (1)	1257	
BPP-2	0.93 (1)	1.07 (1)	0.85 (1)	0.93 (1)	1.09 (1)		0.96 (1)	0.94 (1)	0.94 (1)	0.83 (1)	0.84 (1)	1014	1014.31
BPP-3	0.95 (1)	1.11 (1)	0.90 (1)	1.00 (1)	1.23 (1)		0.95 (1)		1.04 (1)	0.92 (1)	0.98 (1)	913	913.14

Table 2. Amino acid sequence of peptides BPP

Peptide	Sequence
BPP-1	H_2NAla-Pro-Val-Leu-Gln-Ile-Lys-Lys-Thr-Gly-Ser-AsnOH
BPP-2	H_2NAla-Pro-Val-Leu-Gln-Lys-Thr-Gly-Ser-AsnOH
BPP-3	H_2NAla-Pro-Val-Leu-Gln-Lys-Gly-Ser-AsnOH

We have studied the effect of BPP peptides on spleen lymphocytes from mice of the Kunming inbred line. The responsive proliferation to mitogen (i.c. Con-A) was determined with a modified Mosmann method [4]. Lymphocyte transformation was measured by counting the percentage of transformed lymphocytes under microscopy with Geimsa's staining after culturing the lymphocytes with mitogen for 72 hours. The results showed that BPP-1 and BPP-3 at the concentration of 1-10 μg/ml have immuno-enhancing effects on responsive proliferation and lymphocyte transformation compared to control (shown in the Table 3).

Table 3. Effect of peptides BPP on lymphocyte function of mice

Peptide	Concentration (μg/ml)	Responsive Proliferation	Lymphocyte Trans-formation (%)
BPP-1	100	0.290	15
	10	0.325	42
	1	0.331	27
	0.1	0.155	8
BPP-3	100	0.258	17
	10	0.314	35
	1	0.344	44
	0.1	0.248	23
Non-peptide		0.300	3
Control		0.001	1

Acknowledgment

This work was supported by the National Science Foundation of China (No. 29672002).

References

1. Wan, K.K., Song, Y.L. and Jin, S., Int. Chinese Peptide Symposium, Beijing, 1994, p. 116.
2. Merrifieid, R.B., J. Am. Chem. Soc., (1963) 2149.
3. Chang, J.Y., J. Chromatogr., 140 (1977) 125.
4. Tim, Mosmann, J. Immunol. Method, 65 (1983) 55.

Synthesis and structure-activity relationships of tripeptide mimetic analogs of ACE inhibitors

Shi-Feng Liu and Jie-Cheng Xu
Shanghai Institute of Organic Chemistry, Chinese Academy of Sciences,
354 Fen Lin Lu, Shanghai 200032, China

Introduction

Angiotensin converting enzyme (ACE) inhibitors hold great promise in the treatment of hypertension [1]. As more is learned about the design and synthesis of protease inhibitors, the replacement of amide backbone by a structure-like surrogate to impart conformational integrity or resist proteolytic degradation can be a promising approach. e synthesized two types of tripeptide mimetic analogs and explored their structure-activity relationships with aim of identifying highly potent and low side effect ACE inhibitors.

Results and Discussion

Azatripeptides were synthesized via the coupling of *N*-protected α-amino acids with azaalanyl proline *tert*-butyl ester followed by deprotection [2]. α-Lactam-bridge tripeptide isosteres were prepared by coupling *N*-protected phenylalanine with γ-lactam-bridged dipeptides [3]. All synthetic compounds were characterized by H-NMR and MS which showed the expected chemical structures and satisfactory elemental analyses.
In vitro ACE inhibitory activities of synthetic tripeptide analogs were tested by a procedure similar to that described by H. M. Neels et. al. with Hip-Gly-Gly as substrate [4]. The results are shown in Tables 1 and 2. Data in Table 1 show that the inhibitory activities depend on the configuration of the chiral carbon and the nature of R group as defined in the general structure. For example, 1f was active by 18 folds as compared to 1g and 1c was 5 folds more active than 1e owing to the higher acidity of the NH group attached to the α-N which favored hydrogen bonding.

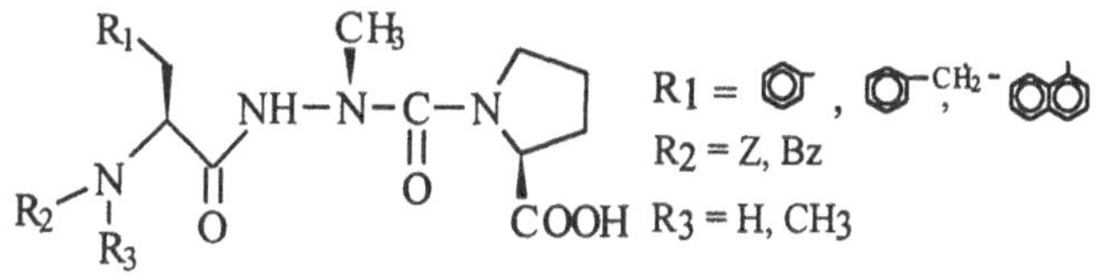

Azatripeptides

Table 1. Inhibitory activities of azatripeptides

No.	R1	R2	R3	Ki(nM)	IC_{50}(nM)
1a	Ph	H	Bz	18	33.3
1b	PhCH2	H	Bz	69	127.9
1c	Ph	H	Z	36	66.7
1d	Ph	H	Z	100	185
1e	Ph	CH3	Z	175	324
1f	1-Nap(S)	H	Bz	28	51.9
1g	1-Nap(R)	H	Bz	>500	>927
captopril				4.6	8.5

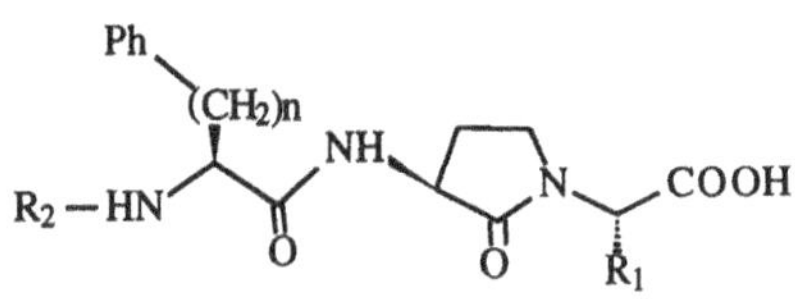

$R_1 = CH_3$, Bn, H
R_2 = Z, Bz
n = 1, 2

γ-Lactam-bridged tripeptides

Table 2. Inhibitory activities of γ-Lactam-bridged tripeptides

No.	R1	R2	n	Ki(nM)	IC_{50}(nM)
2a	H	Z	1	100	185
2b	H	Bz	1	45	83.4
2c	CH3	Bz	1	32	59.3
2d	Bn	Bz	1	52	96.4
2e	H	Z	2	170	315

References

1. Patchett, A.A., Nature (London), 288 (1980) 280.
2. Weinstein, J., European patent application 005486.
3. Freidinger, R.M., J.Org. Chem., 47(1982) 104.
4. Neels, H.M., Clin. Chem., 28 (1982) 1352.

Structure-activity relationship of VIP-PACAP hybrid peptides

Mu-Xin Wei[a, b], S. Naruse[c], K. Nokihara[d], T. Ozaki[b], E. Ando[d], De-Zhen Chen[a], Zhen-He Shun[a] and V. Wray[e]

[a]*Nanjing Medical University, Nanjing 210029, China*
[b]*National Institute for Physiological Sciences, Okazaki 444, Japan*
[c]*The Second Department of Medicine, Nagoya University School of Medicine, Nagoya 466, Japan*
[d]*Central Laboratory, Shimadzu Corp., Kyoto 604, Japan*
[e]*Gesellschaft für Biotechnologische Forschung, Braunschweig D-38124, Germany*

Introduction

Pituitary adenylate cyclase activating polypeptide [PACAP] is a neuropeptide structurally related to vasoactive intestinal polypeptide [VIP]. It exists in the ovine hypothalamus in two molecular forms, PACAP38 and PACAP27 [1].The amino terminal 27 residue sequence of PACAP38 [PACAP27] has high homology (68%) with that of VIP. Many biological actions of PACAP are similar to those of VIP, but we have recently found that PACAP contracts the guinea pig gallbladder whereas VIP relaxes it [2]. This system is a good model for studying the structure-activity relationship of PACAP and VIP. In the present investigation we have synthesized the PACAP-VIP hybrid peptides by a simultaneous multiple solid-phase peptide synthesizer using Fmoc strategy and compared the effect of the peptides on gallbladder smooth muscle *in vitro*.

Results and Discussion

Considering of the sequence homology of VIP-PACAP and results based on the studies by NMR and computer aided molecular graphics [3], positions 4, 5, 6, 8, 9, 13, 21, and 24-26 were selected as mutation points for the present study (Fig. 1). The peptides were tested on isolated guinea pig gallbladder using an improved horizontal-type organ bath described previously [4]. VIP induced relaxation, while PACAP27, on the other hand, induced concentration-dependent contraction. Position 4, 5, 9 and 24-26 can be replaced without significant loss of activity. [Leu^{13}]-PACAP27, a substitution in the α-helix domain, has no significant loss of activity, and is more potent than [Gly^{8}]-, [DAsp^{8}]-PACAP27 and substituted peptides at position 21. Des-[His^{1}] and [Ala$_{6}$]-PACAP27 had no activity. [Gly^{8}]-, [DAsp^{8}]-, [Phe^{21}]- and [Pro^{21}]-PACAP27 at 10^{-7} M were about 25% as active as PACAP27 (10^{-7} M). In our previous studies, the *N*-terminus from position 1 to 8 showed no defined helix or strand structure. Substitution of PACAP in this region showed less potency than substitutions in other regions. These data suggest that the disordered region from 1 to 8 is very important for physiological action. Position 21 is also important, however, because at a higher dose ($3x10^{-7}$M) there was no significant loss in activity.

VIP	: HSDAVFTDNY TRLRKQMAVK KYLNSILN-NH_2
PACAP27	: HSDGIFTDSY SRYRKQMAVK KYLAAVL-NH_2
Des-[His1] PACAP27	: SDGIFTDSY SRYRKQMAVK KYLAAVL-NH_2
[Ala4]-PACAP27	: HSD**A**IFTDSY SRYRKQMAVK KYLAAVL-NH_2
[Val5]-PACAP27	: HSDG**V**FTDSY SRYRKQMAVK KYLAAVL-NH_2
[Ala6]-PACAP27	: HSDGI**A**TDSY SRYRKQMAVK KYLAAVL-NH_2
[Gly8]-PACAP27	: HSDGIFT**G**SY SRYRKQMAVK KYLAAVL-NH_2
[Dasp8]-PACAP27	: HSDGIFT**D**SY SRYRKQMAVK KYLAAVL-NH_2
[Asn9]-PACAP27	: HSDGIFTD**N**Y SRYRKQMAVK KYLAAVL-NH_2
[Leu13]-PACAP27	: HSDGIFTDSY SR**L**RKQMAVK KYLAAVL-NH_2
[Ala21]-PACAP27	: HSDGIFTDSY SRYRKQMAVK **A**YLAAVL-NH_2
[Phe21]-PACAP27	: HSDGIFTDSY SRYRKQMAVK **F**YLAAVL-NH_2
[Pro21]-PACAP27	: HSDGIFTDSY SRYRKQMAVK **P**YLAAVL-NH_2
[Asn24, Ser25, Ile26]-PACAP27	: HSDGIFTDSY SRYRKQMAVK KYL**NSI**L-NH_2

Fig. 1. Amino acid sequence of synthetic peptides.

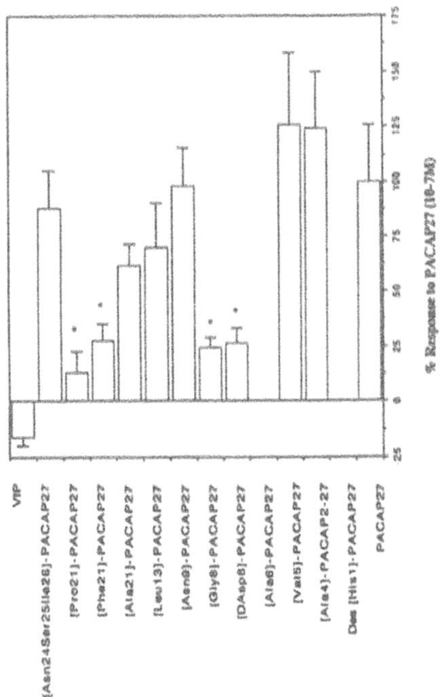

Fig. 2. The effect of PACAP/VIP analogues on isolated guinea pig gallbladder. Mean±SEM are given (n=9 in PACAP27 and VIP, n=5 in PACAP-VIP hybrid peptides); Asterisks indicate significant (P<0.05) differences from PACAP27.

Acknowledgments

This work was supported by Monbusho international scientific research program and a grant from the Ministry of Education, Science and Culture, Japan to S. Naruse. Mu-Xin Wei is a visiting scientist from Nanjing Medical University.

References

1. Miyata, A., Ariimura, A., Dahl, R. R., Minamino, N., Uehara, A., Jiang, L., Cullar, M. and Coy, D.H., Biochem. Biophys. Res. Commun., 164 (1989) 567.
2. Wei, M.X., Naruse, S., Nakamura, T., Nokihara, K. and Ozaki, T., Biomed. Res., 15 (1994) 221.
3. Wray, V., Kakoschke, C., Nokihara, K. and Nakamura, S., Biochemistry, 32 (1993) 5832.
4. Saji, T. and Ozaki, T., In Naruse, S. and Nokihara, K (Eds.) Proceedings of the Workshop for Biosynthesis and Physiological Actions of Vasoactive Peptides, 1992, p.12.

Design and synthesis of salmon calcitonin analogs

Bin Yang, Zong-Jin Han, Ke-Liang Liu and Zhen-Kai Ding
Beijing Institute of Pharmacology & Toxicology, Beijing 100850, China

Introduction

Calcitonin is a peptide hormone with 32 residues that is known mainly for its hypocalcemic effect and inhibition of bone resorption. Although there are many differences in individual amino acids in calcitonins from different species, they all have a disulfide bridge between cysteine residues at positions 1 and 7, and the *C*-terminus is always a proline amide [1]. In general, fish calcitonins have been found to be more potent than mammalian ones. Salmon calcitonin (sCT), which exhibits the highest potency of the native calcitonins, is about 30-fold more potent than human calcitonin (hCT). Native calcitonins and eel calcitonin (eCT) analog are used for treatments of hypercalcemia, osteoporosis and Paget's disease etc. However, the disulfide bonding occurring in each of the natural calcitonins is quite unstable in solution. Numerous CT analogs with the disulfide bridge replaced by other kinds of ring have been reported previously. Although the *N*-terminal disulfide bridge is essential for the biological activity of human calcitonin (hCT), it is not required for that of eCT or sCT [2]. The *C*-terminal proline amide is a very important residue for calcitonins, if it is hydrolyzed to proline, the potency become very weak. Studies have also demonstrated the requirement of the entire peptide structure for high agonist potency, although some residue deletions, e.g., 19-22 segment deletion, are well tolerated [3].

Results and Discussion

Analogs of sCT were synthesized to incorporate modifications which would increase the hypocalcemic effect and improve stability. Structure modifications involved: 1) replacements of Cys residues at positions 1 and 7 with residues Val and Ala respectively to yield open chain analogs, 2) replacements of residue Gly30 with residues Ala, D-Ala and Sar, 3) deletion of segment 19-22 with the above modification, 4) the amide in *C*-terminal proline amide substituted by diethylamide and residue Thr31 replaced by residue Glu.
For the solid phase synthesis of peptides with *C*-terminal proline amide (peptide 1-8), *p*-methylbenzhydrylamine(MBHA) resin and Fmoc / tBu chemistry were employed. Before the cleavage of peptide from the resin, the side chains were first deprotected by treating the fully protected peptide resin with the mixed solvent of TFA / TMSBr / thioanisole (8:1.5:2). Then the peptide was cleaved from the resin with HF, similar to the two-step "low and high" HF treatment in Boc /Bzl chemistry. The formation of disulfide bridge between Cys residues at positions 1 and 7 was carried out on the resin by air oxidation. For the synthesis of the peptide with the *C*-terminal proline diethylamide (peptide 9), the segment Fmoc-Glu(-COOH)-Pro-N(C_2H_5)$_2$ was first reacted with Wang's resin through the γ-carboxyl group of residue Glu followed by stepwise solid phase synthesis. Peptides purified by gel filtration on a Sephadax G-25 column and semi-preparative HPLC were characterized by HPLC (>99% purity) and FAB-MS.

Table 1. Structure, FAB-MS and HPLC retention time (RT) of the peptides*

No.	Amino acid sequences	MW (Calc./FAB-MS)	RT (min)
1	VSNLSTAVLGKLSQELHKLQTYPRTNTGSGTP-NH_2	3397.8 / 3398.5	24.21
2	VSNLSTAVLGKLSQELHKLQTYPRTNTGSATP-NH_2	3411.8 / 3412.3	24.26
3	VSNLSTAVLGKLSQELHKLQTYPRTNTGSSarTP-NH_2	3411.8 / 3410.6	24.01
4	CSNLSTCVLGKLSQELHKLQTYPRTNTGSATP-NH_2	3445.9 / 3446.0	24.64
5	CSNLSTCVLGKLSQELHKLQTYPRTNTGSd-ATP-NH_2	3445.9 / 3446.1	25.16
6	CSNLSTCVLGKLSQELHKLQTYPRTNTGSSarTP-NH_2	3445.9 / 3446.0	25.01
7	VSNLSTAVLGKLSQELHKPRTNTGSGTP-NH_2	2892.2 / 2892.8	22.13
8	CSNLSTCVLGKLSQELHKPRTNTGSATP-NH_2	2940.3 / 2941.0	23.62
9	CSNLSTCVLGKLSQELHKLQTYPRTNTGSGEP-$N(C_2H_5)_2$	3516.0 / 3516.0	25.93

*Mobile phase: A) 70%CH_3CN/0.1%TFA-H_2O, B) 0.1%TFA/H_2O
Gradient: From 100% A to 60% B in 20min, then to 100% A in 10min.
Flow rate: 2.5ml/min. Detection: UV 215nm. Nucleosil C18 column(7μ, 8x250nm).

References

1. Epand, R. M. and Caulfield, M. P., In Emmett, J. (Ed.), Comprehensive Medicinal Chemistry, Vol. 3. Pergamon Press plc, Oxford, 1990, p.1023.
2. Orlowski, R. C., Epand, R. M. and Stafford, A. R., Eur. J. Biochem., 162 (1987) 399.
3. Epand, R. M., Epand, R. F., Stafford, A. R. and Orlowski, R. C., J. Med. Chem., 31 (1988) 1595.

Session III
Conformation of Peptides and Proteins

Chairs: Peter W. Schiller
Clinical Research Institute of Montreal
Montreal, Canada

and

Jie-Cheng Xu
Shanghai Institute of Organic Chemistry
Shanghai, China

Template assembled synthetic peptides (TASP) as receptor mimetics and 'locked-in' tertiary folds

Gabriele Tuchscherer, Pascal Dumy, Patrick Garrouste, Christian Lehmann, Marc Mathieu, Cristina Peggion, Stephane Peluso, Alain Razaname and Manfred Mutter
Institute of Organic Chemistry, University of Lausanne, BCH-Dorigny, CH-1015 Lausanne, Switzerland

Introduction

Protein de novo design aims to mimic some of the structural and functional properties of natural proteins with the ultimate goal of constructing novel macromolecules exhibiting native-like properties. To circumvent the complex protein folding process we have introduced the concept of Template Assembled Synthetic Proteins (TASP), in which templates as 'built-in' device serve to direct the intramolecular assembly of covalently attached peptide blocks into characteristic folding topologies [1]. Due to progress in the design of regioselectively addressable templates [2] and ligation techniques [3] for the condensation of peptide fragments, one of the most challenging applications of the TASP concept, i.e. the construction of TASP molecules acting as functional mimetics of receptors and ligands [4] (Fig. 1), is about to be realized. Conceptually, the functional part of a protein, e.g. the binding loops of a receptor, is detached from the structural part of the molecule and assembled on a topological template which mimicks the structural framework of the native protein. Here, we report on the synthetic methodologies for accessing this novel generation of functional TASP molecules with the example of two prototypes, disposing (i) three loop sequences derived from the phosphorylcholine binding monoclonal antibody McPC603 as receptor mimetic, and (ii) an engineered zinc finger motif exhibiting a 'locked-in' tertiary fold.

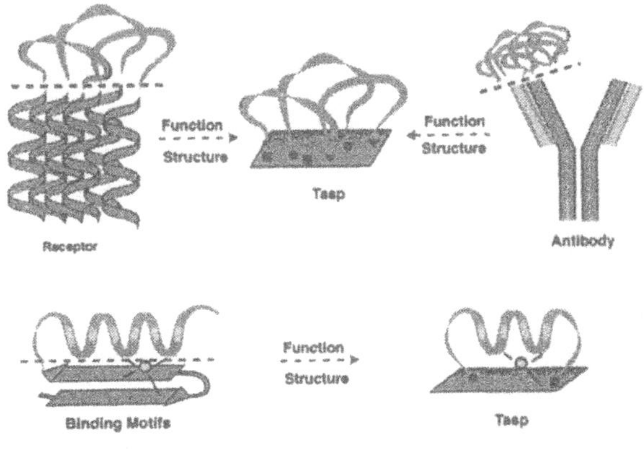

Fig. 1. General concept of separating structure from function in proteins.

Results and Discussion

The general approach (Fig. 1) comprises two key elements, i.e. (i) loop or helical sequences containing *C*- and *N*-terminal functional groups for chemoselective ligation and (ii) topological templates for the regioselective assembly of peptide sequences. Choosing from a variety of orthogonal protecting groups for peptide synthesis, we have developed a series of cyclic peptides as templates for the selective attachment of loop sequences. For example, antiparallel β-sheet peptides of the type $(Xaa\text{-}Ala\text{-}Xaa)_n$ (where Xaa denotes trifunctional amino acids as attachment sites) are linked via β-turn forming dipeptides, turn mimetics or scaffolds; in tailoring the flexibility of the backbone and side chains, this template motif provides any desired number and geometry of attachment sites.

The sequential condensation of peptide sequences onto these topological templates is performed via amide (Scheme 1, strategy A and B), oxime (D) and thioether (C) formation or combinations thereof.

As a first example of a potential ligand (substrate, antigen, transition state molecule) binding TASP molecule, peptides derived from CDR (complementarity determining region) loop sequences of the phosphorylcholine binding monoclonal antibody McPC603 were covalently attached to a topological template.

Starting from a cyclic peptide with selectively addressable pairs of reactive sites (Boc, Dde and Aloc protected lysine side chains), the sequential condensation of three peptides (each with carboxyl groups at the *N*- and *C*- terminus) proceeds smoothly via amide bond formation and provides the target molecule in high yield and purity (Scheme 1, A').

The consensus sequence Xaa_3-Cys-$Xaa_{2\text{-}4}$-Cys-Xaa_{12}His-$Xaa_{3\text{-}4}$-His-Xaa_4 (Xaa representing any amino acid) of zinc finger DNA-binding proteins has been shown to fold in the presence of ZnII into a ββα-folding unit (zinc finger, Zif) with the helical part important

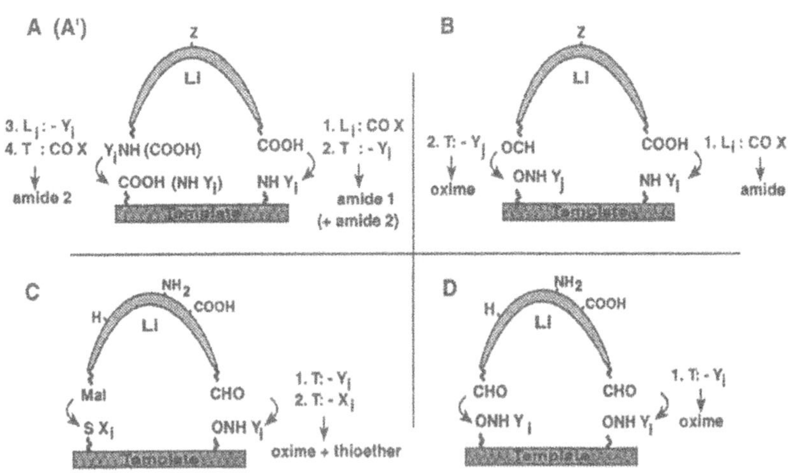

Scheme 1. Different strategies for the attachment of loops to templates; Y_i: protecting groups for NH_2; Z: protection of trifunctional amino acids; X: activation of the carboxylic group.

for DNA binding. Based on the concept of separating structure and function in proteins, a cyclic 14-mer β-sheet template was used as a mimetic of the structural part of the Zif molecule (Fig. 1) to which the helical block was covalently attached via non peptidic linker groups on both chain ends of the helix. The individual building blocks were synthesized, i.e., the template molecule (containing two lysines functionalized with Boc-$NHOCH_2COOH$) and the helical segment (containing serine residues at both chain ends [5]). After simultaneous mild oxidation of the two serine residues with $NaIO_4$, the target molecule was obtained in a single condensation step of the helical block to the template by oxime bond formation (Scheme 1, D).

Preliminary investigations have shown that the Zif-TASP preserves the essential structural and functional features of the native molecule, as shown by UV-, CD- and ^{1}H-NMR-spectra and by ZnII-complexation properties. Contrary to the native linear polypeptide, the TASP molecule retains its structural motif even in the absence of the zinc complex confirming a non native like folding behavior and increased thermodynamic stability.

Conclusion

A new concept for the construction of protein mimetics based on the separation of functional and structural domains in proteins has been evaluated by the successful design and total synthesis of two prototype TASP molecules. This extension of the TASP concept [4] relies on novel synthetic methodologies for the synthesis of topological templates and makes use of the immense repertoire of synthetic organic chemistry for their functionalization. In combination with new protection schemes and ligation techniques it provides the synthetic entry to more complex molecules with a pivotal role in protein and drug design.

Acknowledgments

This work was supported by the Swiss National Science Foundation.

References

1. Mutter, M. and Vuilleumier, S., Angew. Chem., Int. Ed. Engl., 28 (1989) 535.
2. Tuchscherer, G. and Mutter, M., J. Biotechnology, 41 (1995) 197.
3. Nyanguile, O., Mutter, M., Tuchscherer, G., Letters in Peptide Science 1 (1994) 9.
4. Mutter, M., Dumy, P., Garrouste, P., Lehmann, C., Mathieu, M., Peggion, C., Peluso, S., Razaname, S.A. and Tuchscherer, G., Angewandte Chemie, 1996, in press.
5. Vilaseca, L. A., Rose, K., Werlen, R., Meunier, A. and Offord, R. E., Bioconj.Chem. 4 (1993) 515.

Conformational studies of peptides spanning the helical sequence in the molten globule α-lactalbumin

T. Kobayashi, A. Shimizu, M. Ikeguchi and S. Sugai
Department of Bioengineering, Faculty of Engineering, Soka University Hachioji, Tokyo 192, Japan

Introduction

Elucidation of the mechanism of globular protein folding is a major issue in structural biology, and recent progress in the experimental studies on protein folding has improved understanding of transient structural intermediates along the folding pathway. Recent excellent techniques have shown the presence of the intermediate at an early stage of folding, which has the native-like secondary structure in the same regions as those in the native molecule and the destroyed specific tertiary structure. Also, these intermediates have been shown to be identical with the molten globule (MG) state observed for some proteins such as α-lactalbumin (α-LA) as a compact equilibrium unfolding intermediate. Therefore studies on the MG state of α-LA are important for understanding protein folding [1]. The native α-LA is composed of four α-helices (A,B,C,and D) and a β-sheet. Recently it has been shown that in the MG state a native-like backbone topology is partially retained (at least the native B and C helices), and a nonnative hydrophobic core is formed in a region located between the B and C helices in the native state. Therefore, to get further information on stabilization of the MG state, the conformational properties of the peptides that encompass the helices and the hydrophobic core region of α-LA should be investigated in detail. We synthesized four peptides, B14(22-35), C17(84-100), C24(84-107), and H7(101-107), which correspond to the B- and C-helices, the C-helix combined with the hydrophobic core, and only the core region, respectively, in bovine α-LA [2]. Also, to study the folding mechanism of α-LA *in vivo*, interactions of these peptides with a molecular chaperone, GroEL, were studied. The chaperone accelerates in vivo protein folding and inhibits misfolding. It binds to the MG proteins, but the binding mechanism is not yet clear.

Results and Discussion

CD spectra of AcB14 and AcC17 were obtained at 5°C and pH=2, which show the character of the coil (Ac: acetyl group at the N-terminus). The CD spectrum of AcC24 also indicated the coil character, although it differs somewhat from that of AcC17. We compared the CD spectrum of AcC24 with a sum of H7 and AcC17. However, it could not be clearly concluded whether or not AcC24 includes some of the ordered structure. The sequential NOE between the amide protons and the medium-range NOE of ^{1}H-NMR were not observed for C17. The chemical shift deviations from the coil value do not indicate the presence of any ordered structure in C17. The deviations for C24 in aqueous media were also similar to those for C17 in the region 84-100, but the deviations in its C-terminal region (101-107) were significant, as those in H7. Because some of the NOE cross-peaks between the NH resonances were observed in the 101-105 region of C24, local interaction may be expected there. To study interaction between both the helices, the CD spectrum of a mixture of AcB14 and C24 was compared with a sum of the spectra of individual peptides. No structure was found to be induced by interaction between the two peptides. The helices in the MG α-LA seem to be stabilized by their interaction with the other parts of the protein. However, the hydrophobic core region (101-107) may be a possible initiation site of the MG.

Some alcohols (such as trifluoroethanol (TFE)) stabilize the helical structure in proteins. Moreover, in the MG state, hydrophobic media including the alcohols seem to be more desirable than aqueous ones. CD ellipticity of C17 in TFE-H_2O showed the TFE-induced coil to helix transition in a two-state type. However, the conformational transition of C24 could not strictly be in the two-state type. In a range of [TFE] >20%, a gradual decrease in ellipticity was observed. Then 1D and 2D ^{1}H NMR were measured. Its NOESY spectrum at [TFE]=0% has only the d_{NN} cross peaks, but no $d_{\alpha N}(i,i+3)$: C24 has scarcely the helical conformation in aqueous media. The helical NOE cross peaks appeared in aqueous TFE, especially at >15%. Differences in chemical shift of each $C^\alpha H$ in C24 from the coil value were remarkable even at [TFE] <20%. A considerable difference was found even for all of the residues in a sequence of ^{100}G-^{107}H at [TFE]<20%. At [TFE] >20%, some of them were almost constant (such as ^{103}Y), and a remarkable increase in the helical content in a region of ^{100}G-^{107}H cannot be expected at [TFE] >20%. However, for ^{96}L, ^{99}V, ^{100}G,^{104}W and so on, the difference changed considerably at 20% < [TFE]<40%. The chemical shifts of α-methyl protons in ^{96}L changed abnormally by TFE around [TFE]= 20%. To explain such changes in the chemical shifts at [TFE]= 20%, the broken helical form at ^{100}G was introduced as a model of C24,which changes the distance between each $C^\alpha H$ and ^{103}Y, as [TFE] changes, and behaves as a whole extended helix at high [TFE]'s. Moreover, the calculation of the geometric distance achieved by using intensities of the NOESY cross peaks showed that the backbone structure of C24 at [TFE]=20% is more flexible in the ^{100}G-^{107}H region than that at [TFE]=40%. Such a character in a sequence of 100-107 in TFE may give important insights about the MG form of α-LA.

Binding of the peptides, B14 and C17, to GroEL was studied by means of the equilibrium dialysis method at pH=7.2 and 4°C with a dialysis tube (a cut-off MW is about 10^5). The binding constants calculated from differences in the equilibrium concentration of the peptide between outer and inner solutions of the tube after equilibrium of dialysis were 5-10 x10^3/M, which are similar to those for the native α-LA, which hardly binds to GroEL in aqueous media [3]. By comparing also with those, 2-500x10^5/M, of the partially disulfide-reduced α-LA in the MG form, B14 and C17 were concluded to bind hardly to the chaperone GroEL in aqueous media.

References

1. Kuwajima, K., FASEB J., 10 (1996) 102.
2. Shimizu, A., Ikeguchi, M., Kobayashi, T. and Sugai, S., J. Biochem., 119 (1996) 947.
3. Okazaki, A., Ikura, T., Nikaido, K. and Kuwajima, K., Nature Struct. Biol., 1 (1994) 439.

Folding intermediate of β-lactoglobulin with non-native α-helical conformation

Y. Goto[a], D. Hamada[a] and S. Segawa[b]
[a]*Department of Biology, Faculty of Science, Osaka University, Toyonaka, Osaka 560, Japan*
[b]*Department of Physics, Kwansei Gakuin University, Nishinomiya 662, Japan*

Introduction

Protein folding is a process in which a polypeptide chain attains a unique native structure that is determined by its amino acid sequence.Several mechanisms have been proposed to explain the efficient folding process of globular proteins. In particular, a sequential and hierarchical folding mechanism with a molten globule intermediate has been popular [1]. In contrast, a nucleation/condensation model of protein folding has been recently evaluated, in which the rate-limiting step is the formation of a nucleus [2]. In both models, it is generally considered that the intermediate conformational states assume the native-like substructures (hierarchical mechanism of folding).

Bovine β-lactoglobulin is a predominantly β-sheet protein consisting of nine anti-parallel β-strands and one α-helix. It consists of 162 amino acid residues and contains two disulfide bonds as well as one residue of free cysteine. We recently noticed three pieces of evidence indicating that the amino acid sequence of β-lactoglobulin has a markedly high preference for an α-helix conformation [3-5]. First, alcohol, in particular 2,2,2-trifluoroethanol (TFE), induced cooperative transition from the native β-sheet to a highly α-helical structure. Second, the high helical preference was retained even for β-lactoglobulin fragments consisting of 20 amino acid residues. Third, secondary structure predictions indicated that β-lactoglobulin has a surprisingly high α-helical preference. This evidence suggested a case of non-hierarchical folding, in which non-native helical structure is involved in the early stage of folding.To examine the validity of this suggestion, we studied the refolding kinetics of β-lactoglobulin by using stopped-flow circular dichroism (CD).

Results and Discussion

When the unfolded β-lactoglobulin in 3.5 M Gdn-HCl was diluted to the native conditions, the ellipticity at 222 nm exceeded that of the native intensity within the dead time of mixing (overshoot phenomena) and slowly returned to the native intensity. A similar overshoot was also reported by Kuwajima et al. [6]. To characterize the burst-phase intermediate, we measured the wavelength dependence of the refolding kinetics. The overshoot was not limited to around 220 nm but was also found across a broad range of the far-UV region. The far-UV CD spectrum of the burst phase intermediate, constructed by extrapolating the observable phases to time zero, is characteristic for α-helical conformation. TFE stabilizes the α-helical state of β-lactoglobulin and its fragments. Therefore, it was expected that TFE would also stabilize the α-helices of the burst-phase intermediate. As expected, the helical content of the burst-phase intermediate was increased significantly with an increase in the final TFE concentration. We also found that a similar α-helical intermediate is accumulated during the equilibrium unfolding transition induced by Gdn-HCl. On the basis of the TFE and Gdn-HCl-induced equilibrium transitions, we constructed a phase diagram of the native β-sheet, TFE-induced α-helical, and Gdn-HCl-induced unfolded states. The phase diagram indicates that the equilibrium accumulation of the α-helical intermediate is consistent with the high helical propensity of this protein.

The present results indicate that, because of its markedly high helical preference, an α-helical intermediate accumulates during the refolding of β-lactoglobulin. The important questions raised are whether the α-helical intermediate is always required for the formation of the native β-sheets of this protein and whether the α-helical intermediate accelerates or decelerates the folding reaction.

References

1. Kuwajima, K., Proteins: Struct. Funct. Genet., 6 (1989) 87.
2. Fersht, A. R., Proc. Natl. Acad. Sci. USA, 92 (1995) 10869.
3. Shiraki, K., Nishikawa, K. and Goto, Y., J. Mol. Biol., 245 (1995) 180.
4. Hamada, D., Kuroda, Y., Tanaka, T. and Goto, Y., J. Mol. Biol., 254 (1995) 737.
5. Kuroda, Y., Hamada, D., Tanaka, T. and Goto, Y., Folding & Design, 1 (1996) 255.
6. Kuwajima, K., Yamaya, H., Miwa, S., Sugai, S. and Nagamura, T., FEBS Lett., 221 (1987)115.

DNA topoisomerases II: Evolution of their ATP binding site and design of new inhibitors

L. Assairi
Institut de Genétique et Microbiologie CNRS, URA 1354, Université de Paris-sud, Centre d'Orsay, 91405 Orsay Cédex, France

Introduction

DNA topoisomerases II are essential enzymes that are involved in the segregation of chromosomes [1]. All DNA topoisomerases II are dimeric enzymes. However, although bacterial monomers are composed of two subunits eukaryotic monomers are composed of one subunit. Several genes coding for DNA topoisomerases II have now been sequenced from bacteria to eukaryotes and one can observe a strong homology between these genes [2]. However, the genes coding for DNA topoisomerases II have structurally evolved in several steps from two separate genes each one coding for a specific protein in *E. coli* [3]. This evalution proceeded to two adjacent genes each one coding for a specific protein in *B. subtilis* [4] to two adjacent but co-transcribed genes which translated into two specific proteins in *Haloferax* (an archaebacterium living in high salt concentration) [5], and finally to only one gene in eukaryotes [6]. This single gene corresponds to the fusion of the two previous bacterial genes. Variations have been observed concerning the ATP hydrolysis that is necessary for all enzymatic reactions catalysed by DNA topoisomerases II. Indeed, ADP is the product formed upon ATP hydrolysis and inhibits ATP hydrolysis, however, AMP and adenosine inhibit only bacterial DNA topoisomerases II. In addition, coumarin-like novobiocin and coumermycin Al inhibit only bacterial DNA topoisomerases II. The biochemical variation observed must be the consequence of the fusion between the ATPase moiety and the nuclease moiety comprising the whole enzymatic activity.

Results and Discussion

Stable complexes between ATP and DNA topoisomerases II have been obtained after ultraviolet irradiation treatment of archaebacterial [7] and eukaryotic [8] DNA topoisomerases II. The formation of these covalent complexes is inhibited by ADP, the product of the reaction that usually inhibits the ATP hydrolysis catalysed by DNA topoisomerases II during DNA conformational change (relaxation of supercoiled DNA). But AMP and adenosine as well as coumarin (novobiocin, coumermycin Al) inhibit the formation of the [ATP-topoisomerase II] complex specifically at their source. Thus, novobiocin inhibits the formation of the (ATPenzyme) complex only in the case of the archaebacterial enzyme. Since coumermycin Al is a stronger inhibitor than novobiocin for bacterial enzymes, several coumermycin Al derivatives have been proposed [9]. An ATP derivative like the 8-azidoadenosine 5'-triphosphate that is modified on the adenine moiety is hydrolysed by the enzyme with less efficiency than ATP although this derivative is able to promote the relaxation of supercoiled DNA catalysed by DNA topoisomerase II. Other ATP derivatives are also now proposed. These ATP derivatives should enable us to understand the ATP hydrolysis reaction as well as the chemical interaction between the enzyme and ATP. This

information can be used for the design of new specific inhibitors to the ATP recognition site of eukaryotic DNA topoisomerases II.
A strong specific inhibitor could be used in the future as a biocaptor. This would facilitate measurement of the synthesis as well as the regulation of the synthesis of the enzyme during the cell cycle and the isolation of various phosphorylated forms of the DNA topoisomerase II in order to analyse their enzymatic potentialities.

References

1. Holm, C., Goto, T., Wang, J. and Botstein, D., Cell, 41 (1985) 553.
2. Wyckoff, E., Natalie, D., Nolan, J.-M., Lee, M. and Hsieh, T., J. Mol. Biol., 205 (1989) 1.
3. Swanberg, S. and Wang, J., J. Mol. Biol., 197 (1987) 729.
4. Moriya, S., Ogasawa, N. and Yoshikawa, H., Nucl. Acids Res., 13 (1985) 2251.
5. Holmes, M. and Dyall-Smith, M., J. Bact., 173 (1991) 642.
6. Giaever, G., Lynn, R., Goto, T. and Wang, J., J. Mol. Biol., 261 (1986) 12448.
7. Assairi, L., Biochim. Biophys. Acta, 1219 (1994) 107.
8. Assairi, L., Prot. Engin., 9 (1996) 45.
9. Assairi, L., Lett. Pept. Sci., 2 (1995) 169.

Analysis of peptide composition and structure of cytochrome P-450nor2

De-Li Liu[a] **and Hirofumi Shoun**[b]
[a]*College of Life Sciences, Huazhong Normal University, Wuhan 430070, China*
[b]*Department of Applied Biochemistry, Tsukuba University, Japan*

Introduction

Cytochrome P-450 has mono-oxygenase activities, and catalyzes mono-oxygenation reaction. It also plays important roles in detoxication and synthesis of steriod hormone [1-2]. Recently, it was found that a unique cytochrome P-450 was involved in fungal denitrification, acting as a nitric oxide reductase (named P-450nor) [3]. There are 3 kinds of cytochrome P-450nor. One is from *Fusarium oxysporum*, i.e .F. P-450nor. The others existed in *Cylindrocarpon tonkinense*, i. e. C. P-450nor1 and C. P-450nor2. We successfully cloned and sequenced C. P-450nor2 cDNA. The peptide sequence deduced from the cDNA was analysed by using GENETYX software.

Results and Discussion

Cytochrome P-450nor2 catalyzes the reduction of NO to N_2O using NADH or NADPH as direct electron donor [4]. From P-450nor2 cDNA sequence, the coded peptide sequence was predicted (Fig.1). P-450nor2 cDNA coded 408 amino acid residues. Among them, the hydrophobic amino acids were 216 (accounted for 52. 94 %) ; neutral amino acids were 72 (17. 65 %); hydrophilic amino acids were 120 (29. 41%). Cytochrome P-450nors are soluble proteins. This is different from that of membrane-bound P-450s. Microsomal membrane-bound P-450s are expected to contain a signal peptide-like hydrophobic domain in their NH_2 terminus region which serves to anchor these P-450s in the endoplasmic reticulum membrane [5]. We analysed the hydropathy profile for the first 100 residues from

MHATEDETTT	IPRFPFQRAS	AFEPPAEFAR	LRANEPISYV	40
ELFDGSLAWL	VVKHEDVCRV	ATDERLSKER	TRLGFPELSA	80
GGKAAAKNKP	TFVDMDAPAH	MNQRSMVEPF	FTEDHVENLR	120
PYIKETVQGL	LNDMVANGCE	EPVDLIEKFA	LPVPSYIIYT	160
ILGVPFEDLE	YLTEQNAIRS	NGSGTAQEAA	AANQQLLKYL	200
AKLVDQRLQE	PKDDLIGRLV	DQQLVPGHIE	KSDAVQIAFL	240
LLVAGNATMV	NMIALGVVTL	MQNPSQLEEL	KADPTLVPGF	280
VEELCRYHTG	SSMAMKRVAK	EDMELGGKLI	RAGEGIIASN	320
QSANRDEDVF	PNPDVFDMHR	DFDSRDGLGF	GFGPHRCIAE	360
LLAKAELEIV	FETLFATLPD	LRVSIPLDEI	ECTPRHKDVG	400
IVRLPVKW				

Fig. 1. Amino acid sequence of cytochrome P-450nor2.

the NH_2 terminus of P-450nor2. It seems to contain a hydrophilic domain in the NH_2 terminus region. From the primary structure, the second structure is predicted. There are 8 α-helix regions located in 14-32, 96-118, 124-147, 165-179, 185-209, 265-274, 291-321, and 356-380. There are 4 β-folding regions (i. e. 37-62, 149-164, 215-226, and 234-264). There are supersecondary structures βαβ and ααα located in 140-240 region. These structures may be related to the function of P-450nor2. The NADPH binding-site of P-450nor2 is probably located in NH_2 terminus region among the first 164 amino acid residues, but the exact location needs further confirmation.

References

1. Lu, A. Y.H. and West, S. ., Pharmacol. Rev., 31 (1980) 227.
2. Barnes, H.J., Arlotto, M..P. and Waterman, M. .R., Proc. Natl. Acad. Sci. USA, 88 (1991) 5597.
3. Nakahara, K., Tanimoto, T., Hatano, K., Usuda, K. and Shoun, H.J., Biol. Chem., 268 (1993) 8350.
4. Usuda, K., Toritsuca, N., Matsuo, Y., Kim, D.H. and Shoun, H., Appl. and Envir. Microbid,. 61 (1995) 883.
5. Walter, P., Gilmore, R. and Blobel, G., Cell, 38 (1998) 5.

Amphipathic helical potency and its relationship to the biological activity of synthetic calcitonin derivatives

Hou-Li Jiang[a], Bin Yang[b], Bing-Yu Liu[a], Ye-Mei Fan[a], Shi-Jie Shen[a] and Zhen-Kai Ding[b]

[a]National Center of Biomedical Analysis, Beijing 100850, China

[b]Beijing Institute of Pharmacology and Toxicology, Beijing 100850, China

Introduction

Three structural regions: a seven-residue cyclic segment at the N-terminus formed by a disulfide bond between cysteines at positions 1 and 7, an amphipathic α-helix from residue 8 to the helix-breaking proline at position 23, and a hydrophilic random coil segment from residue 23 to the C-terminal proline amide may be important in the biological activity of calcitonin [1]. Four salmon calcitonin (sCT) analogues were designed, synthesized and their conformational properties were studied by Fourier-transform infrared (FTIR) spectroscopy and circular dichroism (CD).

Results and Discussion

The amino acid sequence of sCT was modified both at the N- and C-terminus to study the effects of Cys^1-Cys^7 disulfide removal, Gly^{30} substitution with D- or L-alanine and C-terminal proliamide replacement with diethylamide. FTIR spectra were scanned on the Bio-Rad FTS-65A spectrophotometer. [Val^1, Ala^7]sCT, [Glu^{31}, C-$N(C_2H_5)_2$]sCT, [Ala^{30}]-sCT and [D-Ala^{30}]sCT all showed FTIR results similar to sCT. The amide I band at 1658 cm^{-1} in trifluoroethanol (TFE) indi-cates α-helical structures while a single strong peak at 1645 cm^{-1} in deuterated water mainly shows random structures. The Fourier deconvoluted amide I bands of the four calcitonin derivatives in TFE (Fig. 1) reveal more structural features as peaks around 1658 cm^{-1} represent α-helical structure, peaks from 1620 to 1639 cm^{-1} indicate β-sheet structure, peaks from 1660 to 1690 cm^{-1} indicate turn structures and peaks around 1645 cm^{-1} are characteristic of random structure [2].

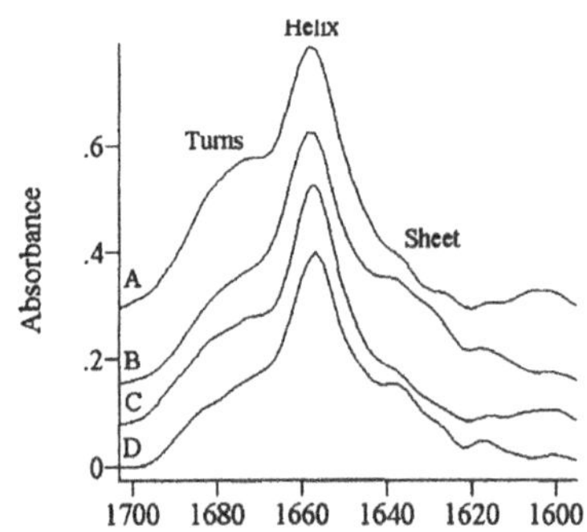

Fig. 1. Deconvoluted amide I spectra of : A. [Val^1, Ala^7]sCT; B. [Glu^{31}, C-$N(C_2H_5)_2$]sCT; C. [Ala^{30}]sCT; and D. [D-Ala^{30}]sCT.

Secondary structure contents measured by FTIR method were obtained by the curve-fitting methods [2]. CD results were calculated by the PROSEC program in the JASCO CD

Table 1. The secondary structures of four calcitonin derivatives in TFE measured by FTIR and CD methods

sCT derivatives	FTIR				CD		
	α	β	T	R	α	β	R
[Val^{1}, Ala^{7}]sCT	48.0	7.9	34.8	9.3	48.6	14.6	36.8
[Glu^{31}, C-$N(C_2H_5)_2$]sCT	45.3	22.2	25.3	7.2	38.3	25.9	35.9
[Ala^{30}]sCT	44.2	9.6	33.3	10.9	40.7	26.1	33.2
[D-Ala^{30}]sCT	48.5	19.9	26.3	5.3	53.7	15.6	30.7

α: α-helix; β: β-sheet; T: turns; R: random structure.

spectropolarimeter. The quantitative results in TFE are shown in Table 1.

The FTIR and CD results are generally consistent only when the random structure measured by CD method includes turns. The study showed that the FTIR methods can provide secondary structural features in more detail. The results indicate that the deletion of disulfide bond at the N-terminus and modification on the C-terminus of natural sCT do not alter the amphipathic topology of the central calcitonin segment, but other regions of calcitonin have significant structural changes.

[D-Ala^{30}]sCT has the highest helical content as revealed by both the FTIR and CD methods. Comparative FTIR studies of [D-Ala^{30}]sCT and sCT in 1,2-propanediol indicate that D-alanine replacement of Gly^{30} changes the sCT β-turn structures at 1674 cm^{-1}, while the contents of α-helix, β-sheet and random coils remain the same.

Hypocalcemic studies showed that the four sCT analogues were all active, and high helical content showed no increase of activity. The fact that [Ala^{30}]sCT has a higher biological activity than [D-Ala^{30}]sCT may indicate that the molecular flexibility is very important for receptor recognition. Although all the four sCT derivatives showed high amphipathic helical potency, the β-sheet, turns and even random coil structures adopted by the N- and C-terminal residues may play an important role.

References

1. Moe, G. R. and Kaiser, E. T., Biochemistry, 24 (1985) 1971.
2. Jiang, H. L., Song, Z. J., Ling, M. S., Yang, S. C. and Du, Z. H., Biochim. Biophys. Acta, 1294 (1996) 121.

DNA-binding characteristics of two analogs of actinomycin D: 5,5'-Val$_2$-AMD and 2,2'-Phe$_2$-AMD

Rui Wang[a], Jing-Man Ni[b], Xiao-Wu Yang[a], Xiao-Yu Hu[a] and Xin-Fu Pan[b]
[a]Department of Biology and [b]Department of Chemistry, State Key Laboratory of Applied Organic Chemistry, Lanzhou University, Lanzhou 730000, China

Introduction

The antitumor activity of actinomycin D (AMD) has been attributed to its interaction with DNA and the most striking biological action is its selective inhibition of DNA-directed RNA synthesis [1]. In order to investigate the DNA-binding ability of the new analogs: 5,5'-Val$_2$-AMD and 2,2'-Phe$_2$-AMD, we applied CD spectroscopy and UV spectroscopy to study the DNA-binding characteristics of the analogs in comparison with those of AMD [2].

Results and Discussion

The UV spectra (Fig.1) and the CD spectra (Fig.2) of the complexes were obtained on UV-260 spectrophotometer and J-20C spectropolarimeter by subtracting the absorption of the drugs.

In Fig. 1. 5,5'-Val$_2$-AMD shows increased the intensity of the A_{260} peak of the DNA (hyperchromic effect) and 2,2'-Phe$_2$-AMD shows decreased value of the A_{260} peak (hypochromic effect). Hyperchromic effect indicated that the drug partially unwinded the

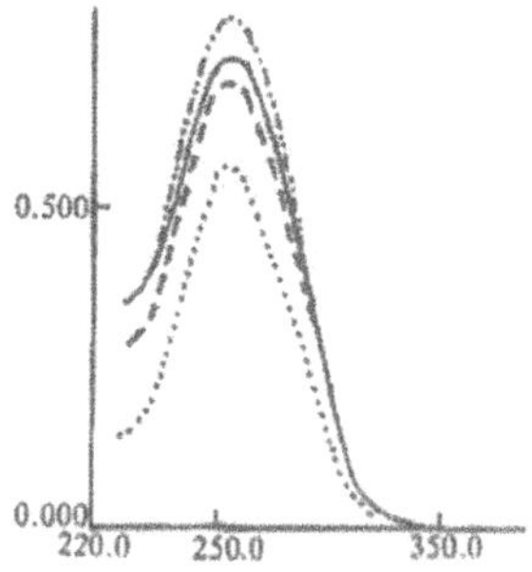

Fig. 1. The UV spectra of complexes of DNA of AMD on its analogs. The solutions (phosphate buffer PH7.2) contained 50 μg/ml drugs: 5,5'-Val$_2$-AMD (-..-), 2,2'-Phe$_2$-AMD(...) and zero μg/ml drugs(—).

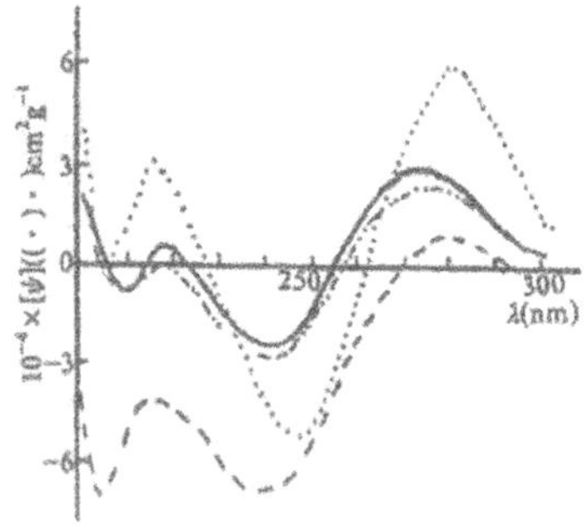

Fig. 2. The CD spectra of complexes of DNA of AMD on its analogs. The solutions (phosphate buffer PH7.2) contained 50 μg/ml drugs: 5,5'-Val$_2$-AMD (-..-), 2,2'-Phe$_2$-AMD(...) and zero μg/ml drugs(—).

double-helix of DNA and hypochromic effect indicated that the drug shrunk it [3]. As shown in Fig. 2, the positive band at longer 273nm of the 5,5'-Phe_2-AMD-DNA complex in CD spectra was reduced and shifted to the wavelength region (red shift) indicating that 5,5'-Val^2-AMD partially unwinded the double-helix of DNA [3]. Both of 2,2'-Phe_2-AMD-DNA complex and AMD-DNA complex showed two unusual ψ-CD spectra, suggesting that the DNA molecules were shrunk [4].

From the above results we concluded that the two peptide analogs could bind to DNA and change the conformation of DNA.

Acknowledgments

This work was supported by grants from the National Natural Science Foundation of China (NO. 29442010, NO. 29502007), Fok Ying Tung Education Foundation and the State Education Commission of China.

References

1. Goldbery, J. H. and Friedman, P. A., Annu. Rev. Biochem., 40 (1971) 775.
2. Dalgleish, D. G., Fey, G. and Kereten, W., Biopolymer, 13 (1974) 1757.
3. Guo, M. L., Yang, P., Yang, P.S. and Zhang, Z. G., Science in China (Ser.B), 5 (1995) 517.
4. Lu, Z. X., In Lu, Z.X. (Ed) Application of Circular Dichroism and Optical Rotatory Dispersion on Molecular Biology, Science Press, Beijing, 1987.

A modified soft docking method for proteins and peptides

Jun-Mei Wang, Xiao-Jie Xu and Fan Jiang
Department of Chemistry, Beida-Jiuyuan Molecular Design Laboratory, Peking University, Beijing 100871, China

Introduction

The computational procedures of docking for proteins and peptides can be classified into three levels by the degree of approximations [1]. There are RBD (rigid-body docking) [2], SFD (semiflexible docking) [3] and FD (flexible docking) [4]. The RBD computation usually uses the crystal structures of bound and unbound complexes to perform docking calculation, while it is useful just for bound complex systems. It is difficult to find the appropriate associated site for unbound complex systems since this model does not allow the conformational change during the docking, however it is common that minor conformational changes are caused when an active molecule associates with its substrate. Some soft-docking procedures have been found in order to overcome the shortcoming of the RBD. One of the successful methods is Fang Jiang 's soft-docking procedure [2].

This procedure uses a 'cube' representation of molecular surface and volume. It designes a simple algorithm of a six-dimensional search to produced implicitly the effects of the conformational changes caused by complex formation. The procedure of the soft-docking may be divided into two stages: the first is the selection of a population of complexes by geometric complementarity. The second is a screening process to identify a subpopulation with favorable energetic interactions. The geometric complementarity is evaluated by the score of the matched surface dots minus the number of atomic pairs which are located in the same cube. The energetic complementarity is evaluated by a quasi-quantitative fashion using 'favor' and 'unfavor' to judge the interactions. This procedure is very useful and effective in finding the appropriate associate site for bounded crystal complexes, however it is not so successful for unbound complex systems.

Methods

In this paper, we developed a procedure based on the idea of the two-stage soft-docking for peptides and proteins with two modifications. First, a new structural parameter is used to screen the unfavorable conformations. The parameter is defined as the number of the atomic pairs whose distances are smaller than the sum of the two Van der Waals radii. If the number exceeds the defined criterion the conformation is discarded. Another modificationis that the second stage of original procedure is replaced by molecular mechanics minimization in six dimensional spaces (three translation freedom and three rotation freedom).We tried to use some widely used minimizing methods such as Steepest descent method, Gauss-Newton method for minimization, but we found they are prone to fall into the potential wells and difficult to escape from them because the potential energy surface is too complicated. As a result one direct search method, Simplex method, has been used in our procedure to overcome the local minimum potential well. The procedure is proved to work well and our computational results are better than Jang's [2].

This program has been written in C. It has been assembled into our molecular design package which can operate under UNIX and DOS operation systems.

The force field parameters are derived from ECEPP. The energetic term is composed of Van der Waals energy, electrostatic energy and H-bond energy between the active molecule and its substrate.

$$\begin{aligned}
&\textit{if } P \leq 3.311 \\
&V_{vanderwaals} = \sum \sum e^{*}(2.9\times10^{5}\exp(-12.5/P) - 2.25P^{6}) \\
&\textit{if } P > 3.311 \\
&V_{vanderwaals} = e^{*}336.176P^{2} \\
&V_{H\text{-}bond} = e^{*}\times r^{6}/R^{12} - 2e^{*}\times r^{5}/R^{10} \\
&V_{electrostatic} = q_i q_j/\varepsilon R
\end{aligned}$$

where r is the sum of radii of atom i and atom j, e* is the depth of the energy minimum, R is the effective internuclear distance, P equal to r/R, q_i and q_j are the partial atomic charges, ε is the dielectric factor.

Results and Discussion

Some protein-protein systems including two bound protein complexes and two unbound protein complexes have been selected to test our program. Table 1 lists the parameters used in our calculation and Table 2 lists the results of the soft-docking calculation. The rms of crystal structures and the calculated conformations for bound complex system are calculated to evaluate the results. Our rms is smaller than that of the original inal procedure [2]. Another advantage progress is that our program can predict unbound complex systems with a quite good result. Rmd (root mean distance) was defined to test the results of

Table 1. Parameters used in the calculation

Molecule names	Number of atoms	Probe radius (Å)[a]	Number of dots	Distance restriction parameter[b]
2PTC[5]	2083	1.7	10524	1/5000
2PTCI/2PTCE	2083	1.7	10524	1/2000
2WRP[6]	843	1.7	5731	1/5000
3B5C/5CYC[7-8]	1496	1.7	8633	1/2000

a The parameter is derived from reference [2].

b The number of atomic pairs whose distances are smaller than the sum of the two Van der Waals radii is defined as a new parameter in our calculation. The criteria of discarding the unfavorable conformation is obtained by the value of distance restriction parameter in Table 1 times the numbers of atoms of active molecule and its substrate.

unbound complex systems. The rmd of system 2PTCI/2PTCE is only 2.19 without optimization with full potential energy field. We are confident that after full minimization the result will be greatly improved.

Table 2. Result of soft-docking calculation

Protein Data Bank code	Solution No.	Rotation (deg)			Translation (Å)			Score[c]	Energy (kJ/mol)	rms/rmd[d]
2PTC	Expt.	0.0	0.0	0.0	3.0	23.2	-10.5			0.05
	1	0.3	0.2	-0.4	3.0	23.0	-10.4	-13.1	-8754.8	
2PTCI/2PTCE	1	101.0	110.0	69.4	14.6	6.9	11.7	-843.4	-7710.7	2.58
	2	80.2	80.2	120.0	8.2	14.9	16.2	-686.6		2.19
2WRP	Expt.	0.0	0.0	0.0	1.3	6.0	-2.9			
	1	0.2	0.4	-0.3	2.1	6.1	-2.2	-17.1	-447.1	0.03
3BIC/5CYC	1	60.4	9.9	151.3	14.4	8.0	-11.1		-15721.4	
	2	92.3	8.6	139.9	5.0	6.7	-5.9		-14518.9	
	3	100.3	20.8	110.2	13.5	9.7	-12.5		-12752.5	
	4	90.0	11.2	119.5	8.8	11.6	-14.8		-12512.9	
	5	59.9	10.9	139.4	13.0	9.1	-11.6		-11043.6	

c The parameter is used to evaluate the geometric complementarity, it is obtained from the number of matched surface dots minus the number of overlapped cubes [2].
d rmd means root mean distance. All the distances of atomic pairs of active molecule and its substrate have been calculated for both bound and unbound complexes. The root mean of the differences of relative distances (rmd) of bound and unbound complexes is used to evaluate the effectivity of the soft-docking calculation. The smaller the rmd, the better conformation have been predicted.

Acknowledgment

This work is supported by National Foundation of Nature Science and National Education Committee Foundation .

References

1. Fraga, S., Parker, J.M., Pocock, J.M., Computer Simulations of Protein Structures and Interactions, Springe-Verlag Berlin Heidelberg, New York, 1995, p. 208.
2. Jiang, F., Kim, S.H., J. Mol. Biol. 21 (1991) 79.
3. Shoichet, K., Kuntz, I.D., J. Mol. Biol. 22 (1991) 327.
4. Hart, T.N., Read, R.J., Proteins Struct. Funct. and Genet. 13 (1992) 206.
5. Marquart, M., Water, J., Deizerthofer, J., Bode, W., Huber, R., Acta Crystallogr., Sect. B,39 (1983) 480.
6. Schevitz, R.W., Otwinowski, Z., Joachimiak, A., Lawson, C.L., Nature, 317 (1985) 782.
7. Mathews, F.S., Argos, P., Levine, M., Cold Spring Harbor Symp., 36 (1972) 387.
8. Takano, T., Dickerson, R.E., J. Mol. Biol., 153 (1981) 79.

Preliminary studies on venom of a marine snail *Conus betulinus*

Ji-Sheng Chen[a], Bing-Gen Ru[b], Fu-Sheng Lin[a], Dong-Mei Yang[c], Pai-Ai Zhou[c], Cai-Hong Wu[c], Shou-Lan Chang[a], Yong-Can Zhou[d], Jian-Ke Tie[b] and Ke-Ping Hu[a]

[a]Research Institute of Pharmaceutical Chemistry Beijing, Beijing 102205, China

[b]National Laboratory of Protein Engineering, Peking University, Beijing 100871 China

[c]National Laboratory of Biomembranes and Membrane Biotechnology, Peking University, Beijing 100871, China

[d]South China Sea Institute of Oceanology, Academia Sinica, Guangzhou 510301, China

Introduction

The *Conus* snails are venomous predators which immobilize their prey by a sophisticated venom apparatus. Hundreds of peptides known as conotoxins exist in these crude venom. Conotoxins provide a series of small peptides which are high-affinity antagonists for receptor and ion channel blockers which are useful effects as special reagents in neurobiology researches. The effects of crude venom and its fractions of *Conus betulinus*, collected from South China Sea, were studied using whole cell patch clamp technique in rats hippocampal CA1 pyramidical cells.

Results and Discussion

Sodium and potassium currents are decreased by application of 0.1mg/ml crude venom and its fractions (Fig. 1 & Fig. 2). It is very interesting to note that the crude venom of *Conus betulinus* can effect the potassium current of a vertebrate neuron, which has not been reported.

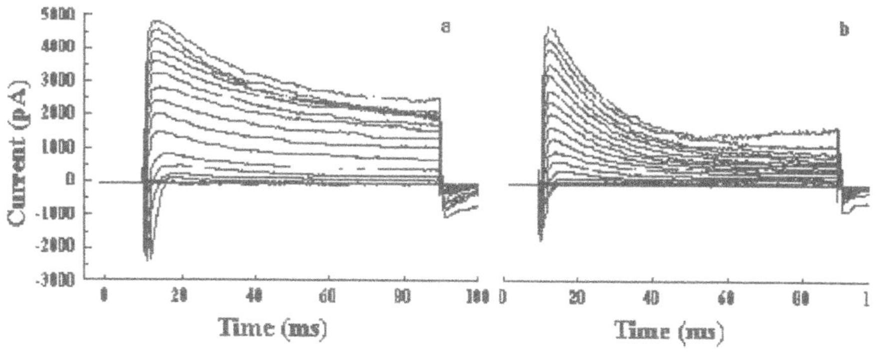

Fig. 1. Effect of crude venom(0.1mg/ml) a: control; b: 6 min after applying venom.

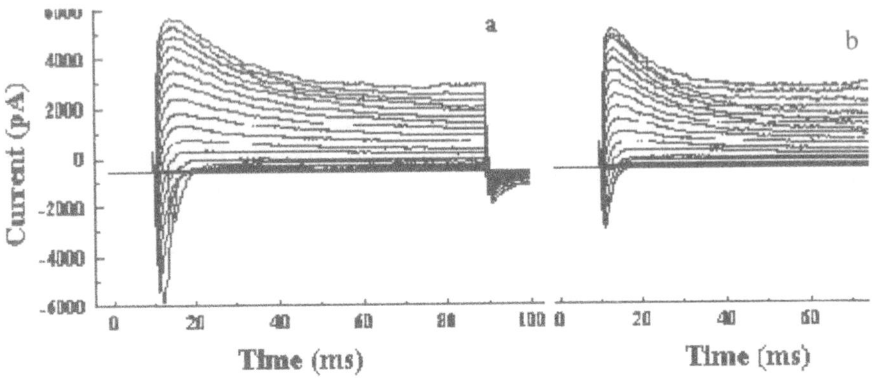

Fig. 2. Effect of peak (0.1mg/ml) a: control; b: 3 min after applying peak E.

Acknowledgment

We thank Pharmacia Beijing Biotech Center in providing a SMART System for our research. We are grateful to Professor Li Lingyaun of Life Science Center, Peking University, Mr Su and Mr Zhu Xi-Can of Pharmacia Biotech for their aids in this work.

References

1. Kohn, A.J., Proc. Natl. Acad. Sci., U.S.A. 42 (1956) 168.
2. Endean, R. and Rudkin, C., Toxicon, 2 (1965) 225.
3. Olivera, B.M., McIntosh, J.M., Cruz, L.J., Luque, F.A. and Gray, W.R., Biochemistry, 23 (1984) 5087.
4. Fainzillber, M., Hasson, A., Oren, R., Burlingame, L.L., Gordon, D., Spira, M.E. and Zlotking, E., Biochemistry, 33 (1994) 9523.
5. McDonough, S.I., Swartz, K.J., Mintz, L.M., Boland, L.M. and Bean, B.P., The Journal of Neuroscience,16 (1996) 2612.

Session IV
Structure and Design

Chairs: James P. Tam
Vanderbilt University
Nashville, Tennessee, U.S.A.

and

Yun-Hua Ye
Peking University
Beijing, China

Topographical considerations in de novo design of peptides

V.J. Hruby[a], S. Liao[a], M. Shenderovich[a], X. Qian[a], G.G. Bonner[a], G. Han[a], E.J. Bilsky[b] and F. Porreca[b]

Departments of [a]Chemistry and [b]Pharmacology, University of Arizona, Tucson, AZ 85721, U.S.A.

Introduction

For *de novo* design of peptide and protein agonists, most investigators have concentrated their efforts on designing or mimicing the peptide backbone secondary structures (ϕ, ψ space) such as α-helix, β-turn and β-sheet structures [1-5]. Using this approach, considerable success has been obtained in the *de novo* design of peptides and proteins with specific secondary structural motifs. In the peptide mimetic area, considerable success has been attained in obtaining non-peptide moieties that can force, mimic or stabilize secondary structures such as α-helices or β-turns [1-3], but most of these structures fail when examined from the viewpoint of the side chain conformational space (topography) since they generally cannot place side chain groups in the same three dimensional space occupied by peptides and proteins. To bridge this gap, and develop a better understanding of the side chain conformational preferences that determine the biological activities and other properties of peptides that depend on peptide side chain groups for their properties, we and others [6-8] have been developing approaches to the design and synthesis of novel amino acids and amino acid mimetics that constrain or fix side chain groups in chi space (chi-1(χ_1), chi-2(χ_2), etc.). We then incorporate these novel amino acids and mimetics into bioactive peptide analogues specifically designed to determine the conformation and topographical three dimensional properties that are critical for biological activity. Examples of such compounds that we have developed and used in biologically active peptides include tetrahydroisoquinoline carboxylic acid (Tic,**1**) [9], β-methyltyrosine [10], β-methyltryptophan [11], and β-methyl-2'6'-dimethyltyrosine (TMT, **2**) [12]. In this paper we report new developments in this area.

Results and Discussion

Figure 1 gives the structures of some of the constrained amino acids that are of interest. In each case the amino acid was designed to mimic an aromatic amino acid but with constraints at χ_1 and χ . In the case of Tic (**1**, Figure 1) the constraint imposed is cyclization, so that only the g (-) and g (+) conformers are possible. We have thoroughly examined this by both NMR [9,13] and X-ray crystallography [13], and have shown its importance in biological structure-activity relationships [9]. Another kind of cyclization is the [2,3] methano analogues of amino acids such as **3**. These amino acids essentially fix the χ_1 torsion angle to give either a cis or a trans relationship of the amine to aromatic moiety. On the other hand, the χ_2 angle is free to rotate constrained only by the Van der Waals steric constraints when 2'- or 6'-methyl groups approach the cyclopropane methylene group. A similar situation is obtained in β-methyl-2',6'-dime-thyl tyrosine (**2**, Fig.1). A

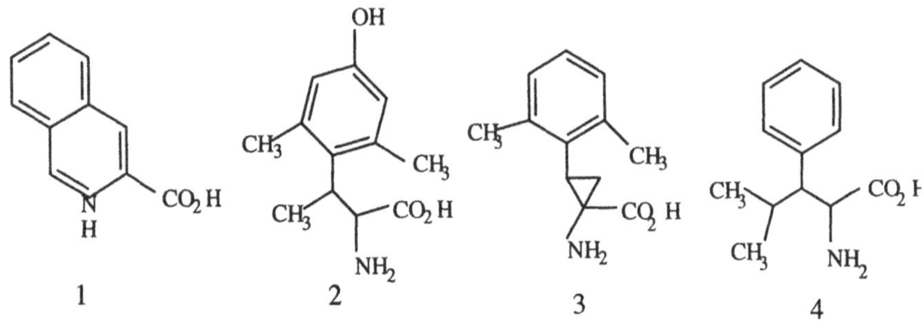

Fig. 1. Structures of topographically constrained amino acids constrained at χ_1 and χ_2

critical issue in **2** and **3** is whether the constraints imposed represent a sufficient constraint to be of biological significance. Thus we have used dynamicNMR methods to determine the barrier to rotation about the Cβ-Cγ bond (chi-2, χ_2). In the case of β-methyl-2',6'-dimethyltyrosine derivatives such as **2**, the barrier was found to be about 16-17 kcal/mole [14], whereas for the 2,3-methano-2',6'-dimethylphenylalanine derivatives such as **3** the barrier was about 15-16 kcal/mole. Thus indeed these compounds are highly χ_2 constrained.

A second critical issue is the extent to which the novel amino acids, though constrained, truly mimic the preferred side chain conformations found for the corresponding natural amino acids. To examine this question we have calculated (χ_1,χ_2) energy maps using the Drive module of MacroModel [15] and for energy calculations, the AMBER force field [16] with a distance dependent dielectric $\varepsilon=4.0_{rij}$. Chi plots of D and L tyrosine and 2 of the 4 isomers of β-methyl-2'6'-dimethyltyrosine are shown in Figure 2. The constrained aminoacids indeed have greatly reduced accessible conformations in chi space with significant differences in energies for g (-), g (+) and trans rotamers in chi-1 space. As for chi-2, again the space available is greatly reduced and the energy barrier calculated is about 17 kcal/mole, consistent with the experimentally determined values. What is most important for applications to biologically active peptides is that the low energy forms in both χ_1 and χ_2 space are similar to those found for the naturally occurring amino acids, except that now only one chi-1 angle is highly favored and there is a high energy barrier between the two preferred χ_2 angles. Thus any effort by the receptor to choose an unfavorable χ_1 and/or χ_2 angle will lead to greatly reduced binding. Indeed we have shown on incorporating all four isomers of **2** into c[D-Pen^2,D-Pen^5]enkephalin (DPDPE), that only one compound, the (2S,3R)-TMT^1 isomer has both high potency and selectivity for the δ opioid receptor [17]. When all four isomers of **2** were incorporated into DPDPE and Deltorphin and examined in *in vivo* antinociception assays the results were extremely interesting as shown in Table 1. What these results demonstrate is that use of topographically constrained amino acid residues in bioactive peptides can affect not only potency,but also can dramatically control the receptor type and subtype selectivity. As expected [(2S,3R)TMT^1]DPDPE, which previously was shown to be a highly selective ligand for δ vs μ receptors, is shown to be highly selective for the δ_1 but not δ_2 receptor,

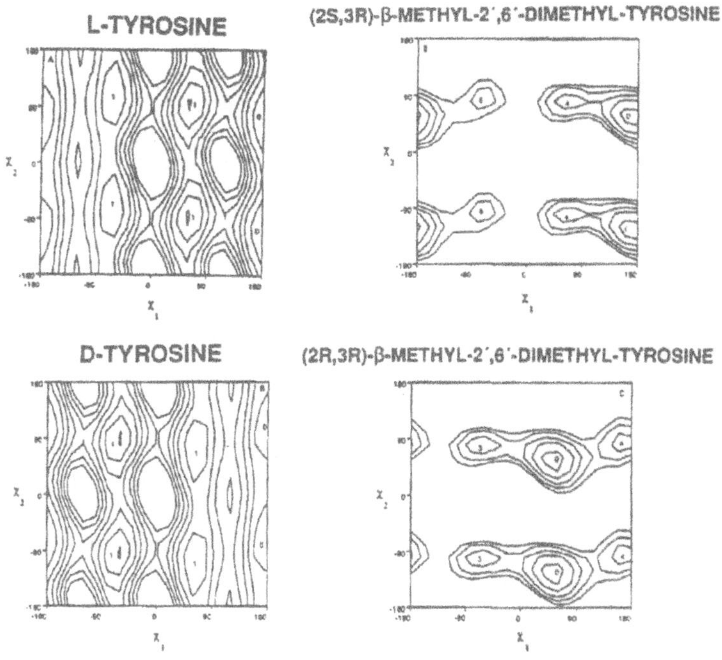

Fig. 2. Chi-1 /Chi-2 energy maps for aromatic amino acids.

just as DPDPE. On the other hand, Deltorphin I, though highly selective for δ vs μ opioid receptors is active at both δ_1 and δ_2 receptors, whereas [(2S,3S)TMT[1]] DELT-I is active only at the δ_2 receptor and [(2R,3S)TMT1]DELT-I is active at both the δ_1 and μ receptors, but not at δ_2 receptors. It is interesting to note that 2R,3S-TMT[1] is a D-amino acid, and yet the analogue is potent at both δ and μ receptors which is contrary to the current dogma in this area.

Table 1. In vivo antinociception activities of analogues

Peptide	Amount used (nM)	Antinociception (%) Following Pretreatment With			
		Control	DALCE (δ1)	[Cys4]DELT (δ2)	β-FNA (μ)
DPDPE	30	89	**30**	82	91
[(2S,3R)TMT[1]]DPDPE	100	90	**42**	72	52
Deltorphin I	3	91	**30**	**33**	78
[(2S,3S)TMT[1]]DELT I	30	87	81	**32**	82
[(2R,3S) TMT[1]]DELT I	30	90	**45**	83	**30**

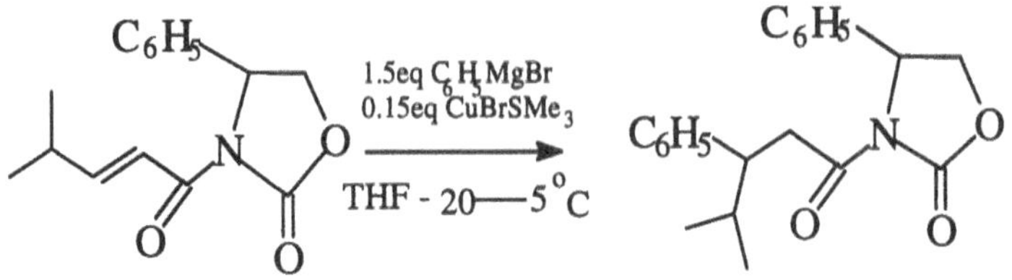

These and other exciting results have prompted us to prepare even more constrained amino acids such as β-isopropylphenylalanine [(**4**, Fig. 1), or β-phenylleucine]. Asymmetric synthesis poses some problems due to steric effects, especially for setting the chirality at the β-carbon. In order to obtain the desired isomer in high enantiomeric excess the special conditions shown provided the desired intermediate **10** in good yield, and then the L and D amino acids can be prepared asymmetrically from **10** utilizing methods already established in our laboratory. It is interesting to note, however, that if the Michael-like addition is done under other conditions, including those previously used in our laboratory for less hindered cases, 1,2-addition becomes the major reaction pathway. We are continuing to develop synthetic methods for these and other highly hindered chimeric amino acids and to incorporate them into bioactive peptides.

Acknowledgment

This work was supported by grants from NIDA DA08657 and DA 06284.

References

1. Rizo, J. and Gierasch, L.M., Ann. Rev. Biochem.., 61 (1992) 387.
2. Giannis, A. and Kolter, T., Angew. Chem. Int. Ed. Engl., 32 (1993) 1244.
3. Farlie, D.P., Abbenante, G. and March, D.R., Curr. Med. Chem., 2 (1995) 654.
4. Hruby, V.J., Life Sci., 31 (1982) 189.
5. Hruby, V.J. and Sharma, S.D., Curr. Opin. Biotech., 2 (1991) 599.
6. Hruby, V.J., Al-Obeidi, F. and Kazmierski, W.M., Biochem. J., 268 (1990) 249.
7. Toniolo, C., Int. J. Peptide Protein Res., 35 (1990) 287.
8. Hruby, V.J., Biopolymers, 3 (1993) 1073.
9. Kazmierski, W.M. and Hruby, V.J., Tetrahedron, 44 (1988) 697.
10. Nicolas, E., Russell, K.C., Knollenberg, J. and Hruby, V.J., J. Org. Chem., (1993) 58.
11. Haskell-Luevano, C., Boteju, L.W., Miwa, H., Dickinson, C., Gantz, I., Yamada, T., Hadley, M.E. and Hruby, V.J., J. Med. Chem., 38 (1995) 4720.
12. Qian, X., Kover, K.E., Shenderovich, M.D., Misicka, A., Zalewska, T., Horvath, R., Davis, P., Porreca, F.,Yamamura, H.I. and Hruby,V.J., J. Med. Chem, 37 (1994) 1746.
13. Valle, G., Kazmierski, W.M., Crisma, M., Bonora, G.M., Toniolo, C. and Hruby, V.J., Int. J. Peptide Protein Res., 40 (1992) 222.
14. Jiao, D., Russell, K.C. and Hruby, V.J., Tetrahedron, 49 (1993) 3511.
15. MacroModel Interactive Molecular Modeling System., Version 4J., Department of Chemstry, Columbia University, NY, 1994.
16. Weiner, S.J., Kollman, P.A., Case, D.A., Singh, U.C., Ghio, C., Alagona, G., Profeta, Jr.S. and Weiner P., J. Am. Chem. Soc., 106 (1984) 765.

17. Qian, X., Shenderovich, M.D., Kover, K.E., Bilsky, E.J., Horvath, R., Zalewska, T., Yamamura, H.I, Porreca, F. and Hruby, V.J., J. Am. Chem. Soc., 1996, in press.

Anticoagulant activity of synthetic peptides targeting various functional steps of blood clotting

Du-Sheng Cheng, Feng-Yun Liu, Xiu-Zhen Li and Guo-Li Wang
Laboratory of Peptide Chemistry, Institute of Biotechnology, Beijing 100071, China

Introduction

Thrombin and fibrin play key roles during blood clotting. Inhibitors of thrombin and fibrin aggregation may be developed to antithrombotic drugs. Hirolog-I is a bifunctional antithrombin peptide combining the catalytic-site antithrombin peptide and exosite antithrombin peptide with a polyglycyl linker [1]. FPRP is an anti-fibrin aggregation peptide [2]. To develop more effective antithrombotic peptides, we designed and synthesized a trifunctional peptide combining Hirolog-I and FPRP with a 4-glycyl linker and compared it with Hirolog-I in antithrombin and anticoagulant activities.

Results and Discussion

Trifunctional peptide(TPI) and Hirolog-I were prepared using multipin polypeptide synthesizer from Australia Chiron Co. Synthetic peptides were purified to homogeneity by preparative C_{18} column employing a Waters HPLC. Anti-thrombin catalytic activities were monitored at 410nm [3] using the Chromozym TH substrate. Anticoagulant activities were determined by observing the fibrin clotting time under different conditions [4]. The results showed that TPI and Hirolog-I had similar antithrombin catalytic activity (Fig. 1). TPI showed evident anticoagulant activity at concentrations of $4x10^{-6}$M TPI at $2x10^{-6}$M showed similar anticoagulant activity of Hirolog-I at $6x10^{-6}$M when thrombin reacted with the peptides (Table 1). When fibrinogen reacted with the samples, TPI expressed good anti-

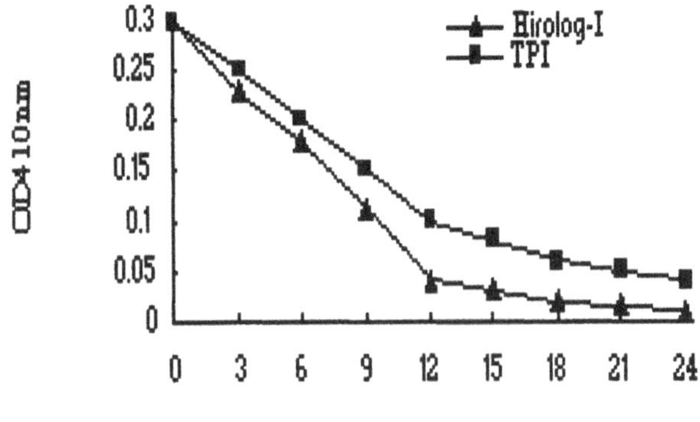

mg/ml ($\times10^{-2}$)

Fig. 1. Comparison of antihrombin catalytic activity of Hirolog-I and TPI.

Table 1. Comparison of antithrombin properties of Hirolog-I and TPI

peptides	concentration ($x10^{-6}M$)	clotting time (s)
Hirolog-I	0	1354
	0.75	2525
	1.5	5726
	3	11014
	6	12527
	0	1402
TPI	0.5	5405
	1	8407
	2	13126
	4	>2100

Table 2. Comparison of anticoagulant properties of Hirolog-I and TPI

peptides	concentration ($x10^{-6}M$)	clotting time (s)
Hirolog-I	0	1302
	0.75	5423
	1.5	70010
	3	84010
	6	102015
	0	1405
TPI	0.5	5605
	1	78010
	2	112015
	4	23010
	80	>2500

coagulant activity at concentration of $8x10^{-6}M$ (Table 2). The above experiments indicated that TPI showed both anti-thrombin and anti-fibrin aggregation activities, suggesting that TPI probably has higher anti-thrombin potency than Hirolog-I and providing an approach to construct multifunctional peptide in future.

References

1. Kelly, A.B., Maraganore, J.M., Bourdon, P., Hanson, S.R. and Harker, L.A., Proc. Natl. Acad. Sci. U.S.A., 89 (1992) 6040.
2. Laudano, A.P. and Doolittle, R.F., Proc. Natl. Acad. Sci. USA, 75 (1978) 3085.
3. Chang, J.Y., FEBS Lett., 164 (1983) 307.
4. Mao, S.J.T., Yates, M.T., Owen, T.J. and Krstenansky, J.L., Biochemistry, 28 (1988) 817.

A rational approach to *de novo* design of α-helical proteins

Zhen-Wei Miao, Xiang-Qun Li, Chong-Xi Li, Xiao-Jie Xu and You-Qi Tang
Department of Chemistry, Peking University, Beijing 100871, China

Introduction

Protein *de novo* design is an inverse but effective approach to elucidate the relationship between protein tertiary structure and its amino acid sequence. Examples include *de novo* designed α-helical coiled coils by Hodges'group [1] and four helical bundles by DeGrado and co-workers [2]. However, *de novo* designed peptides usually show some characteristics of molten globule rather than the expected native-like state [3]. Particularly, some can not fold into their predicted structures [4,5]. Because of the low predictability, we have to systematically investigate the factors that dominate the formation and stability of α-helix structure by designing and characterizing model I peptide in our previous work [6]. In this paper, we will present a rational approach to the helical protein design based on the known structures of leucine zipper peptide GCN4-p1 mutants [7, 8].

Model I: Z-(E-E-X-Y-K-K-X)n-G Z= Ala, Ser; X=Leu, Ile, Val, Phe, Ala; Y=Ala,Leu; n=2,3,4;

Results and Discussion

Harbury et al [7,8] have systematically substituted the hydrophobic residues at positions "a" or "d" of the GwCN4 coiled coil with a single amino acid. The mutants obtained display two-, three-, and four-helix structures depending on the match of the buried residues [7,8]. Accordingly, we have designed LI, IL, and VL peptides with computer modelling:

LI	**C-(E-Id-E-Q-K-La-K)$_n$**	(LI22, n=3; LI29,n=4)
IL	**(E-Ld-E-Q-K-Ia-K)$_n$**	(IL21,n=3; IL28, n=4; IL35, n=5)
VL	**(E-Ld-E-Q-K-Va-K)$_n$**	(VL21, n=3; VL28, n=4; VL35,n=5)

The sequences of LI, IL and VL are similar to each other except for the match of hydrophobic residues at positions a and b. The calculation as well as visual inspection clearly show that the different match of a and b will lead to different aggregation (Fig.1). LI peptides are favorable to fold into parallel four-helix bundles (Fig.2), whereas IL and VL peptides tend to form parallel di-strand coiled coils. The molecular size-exclusive chromatography and circular dichroism (CD) studies confirmed our prediction, as summarized in Table 1. LI22, LI29, IL35 and VL35 can self-assembly into their predicted folds with high content of helix in the neutral PBS or MOPS buffers. However, IL21, IL28, VL21 and VL28 exist in an equilibrium between desired dimer and higher-order oligomers.

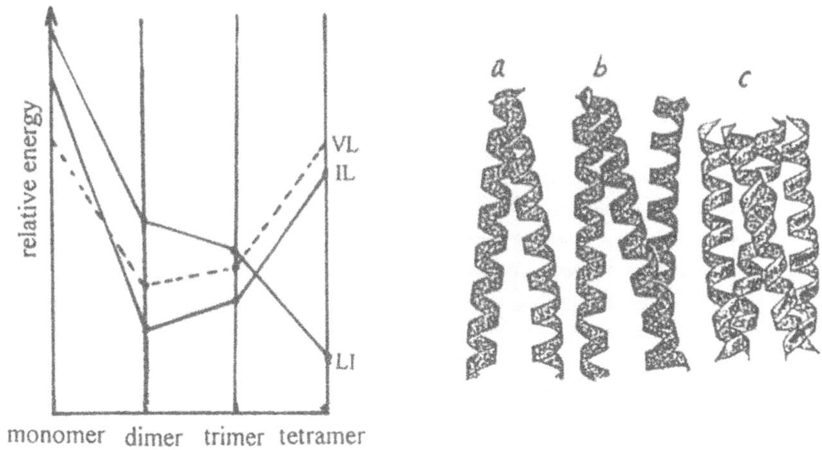

Fig. 1. (A)Relation of peptide oligomers and their packing energy.

(B)Diagram of peptide oligomers (side chains are omitted).

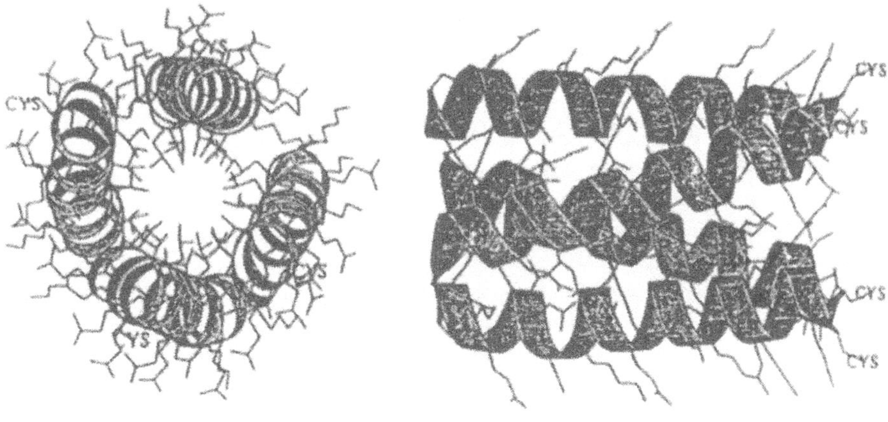

(A) end view

(B) side view

Fig. 2. The molecular models of LI peptides predicted to form parallel four helical bundles.

Further CD studies illustrate that the ellipticity values at 222nm of LI22, LI29, IL35 and VL35 are concentration-independent even diluted to eight folds from 0.27mM to 0.034mM, indicating that these peptides can associate into their uniquely tertiary folds. When 30% TFE is used as a conformation-inducing agent, the helix contents of IL21 and VL21 are increased dramatically. In contrast, the helicities of IL28 and VL28 increase a little, while the long peptides IL35 and VL35 are essentially 100% helical in neutral aqueous buffers compared with those in the presence of 30% TFE.

Table 1. Summary of peptide structures of LI, IL and VL in aqueous solution

peptide	position a	position d	$\theta_{[222]}$ deg·m^2·mol^{-1}	No. of aggregation predicted	No. of aggregation observed
LI22	L	I	31,300	4	4
LI29	L	I	43,200	4	4
IL21	I	L	24,700	2	3, 6
IL28	I	L	38,000	2	2, 4
IL35	I	L	43,300	2	2
VL21	V	L	24,300	2	2,4
VL28	V	L	35,400	2	2,4
VL35	V	L	42,000	2	2

The association stability of LI, IL and VL peptides has been determined by guanidine hydrochloride (Gu·HCl) denaturation and thermal unfolding profiles. As showed in Figs. 3-5, the [Gu·HCl]50 values of peptides vary markedly with chain length and concentration. The stability against guanidine hydrochloride denaturation follows VL28<IL28<LI29 and VL35<LI29<IL35. The thermal unfolding profiles also give a similar result, as demonstrated in Fig. 6. Usually, the stability against thermal unfolding can be compared by their Tm which is the temperature standing for 50% denaturation. However, some Tm can not be obtained from the thermal unfolding profiles directly. Thus, the decrease of ellipticity at 222 nm is selected as a reference when temperature rises from 8°C to 80°C. It could be found that IL29 and IL35 are very stable to thermal melting. Their helicities only reduce 17.2% and

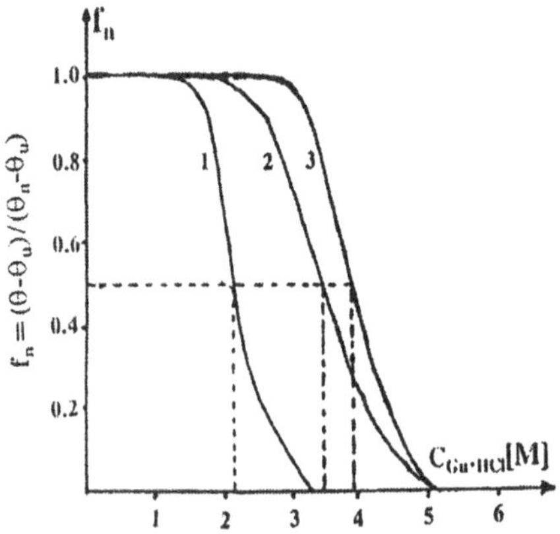

Fig. 3. Guanidine denaturation profiles of LI22 and LI29 1, LI22(0.27mM) 2, LI29(0.08mM) 3, LI29(0.27mM)

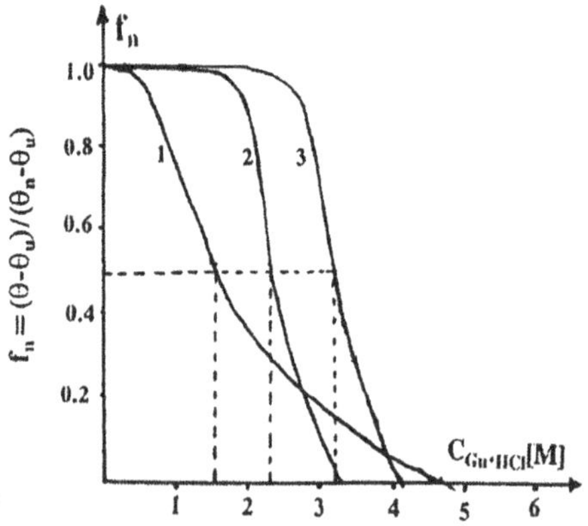

Fig. 4. Guanidine denaturation profiles of VL21, VL28 and VL35 1, VL21(0.27mM) 2, VL28(0.27mM) 3, VL35(0.08mM).

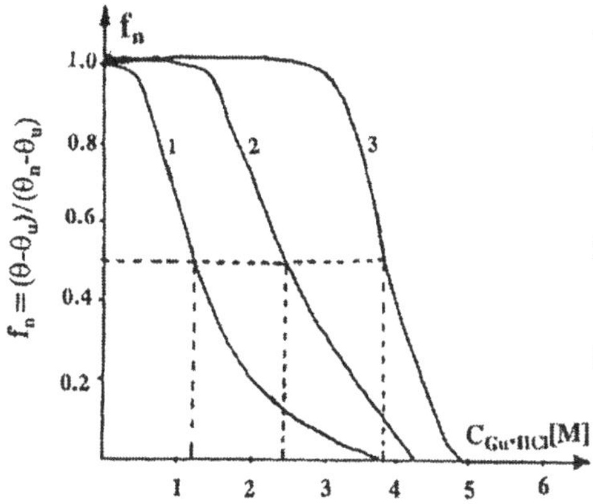

Fig. 5. Guanidine denaturation profiles of IL21, IL28 and IL35 1, IL21(0.27mM) 2, IL28(0.27mM) 3, IL35(0.05mM).

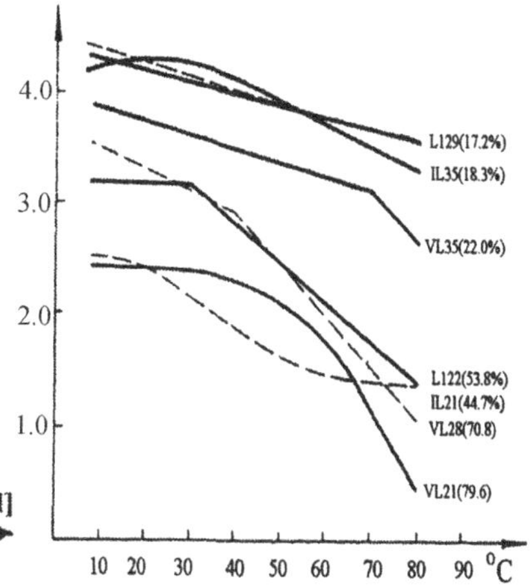

Fig. 6. Thermal melting profiles and percentage of ellipticity decrease.

18.3% respectively when temperature was raised from 8°C to 80°C. In addition, their linear melting curves account for both peptides retaining their unique tertiary folds during thermal denaturation. In summary, the present studies provide a rational approach for designing *de novo* helical proteins with improved predictability.

References

1. a, Lau, S. Y. M., Taneja, A. K., and Hodges, R. S., J. Biol. Chem., 259 (1984) 13253. b, Monera, O. D., Zhou, N. E., Kay, C. M., Hodges, R. S., *ibid*, 268 (1993) 19218.
2. DeGrado, W. F., Wasserman, Z. R. and Lear, J. D., Science, 243 (1989) 622.
3. a, DeGrado,W. F., Raleigh, D. P. and Handel, T., Curr. Opin. Struct. Biol., 1 (1991) 984.
4. b, Betz, S. F., Raleigh, D. P., DeGrado, W. F., *ibid,* 3 (1993) 601.
5. Lovejoy, B. Science, 259 (1993) 1288.
6. Hill, C. P., Anderson, D. H., Wesson, L., DeGrado, W. F. and Eisenberg, D., Science, 249 (1990) 543.
7. a, Xu, X. J., Luo, Y., Zhang, R. H. and Miao, Z. W. In Ye, Y. H. In Lu, G. S. and Tam,J.P., Du,Y. C. (Eds) Peptides: Biology and Chemistry (Proceedings of the 1994 Chinese Peptide Symposium), ESCOM, Leiden, 1995, p.97. b, Li, X. Q., Miao, Z. W., Luo, Y., Xu, X. J. and Tang, Y. Q., Chinese Chem. Lett., 7 (1996) 812.
8. Harbury, P. B., Zhang, T., Kim, P. S. and Alber, T., Science, 262 (1993) 1401.
9. Harbury, P. B., Kim, P. S. and Alber, T., Nature, 371 (1994) 80.

Synthesis of a series of amino acid and peptide-porphyrin derivatives

Xiao-Hua Chen, Xiao-Lin Hao, Gui-Ling Tian and Yun-Hua Ye
Department of Chemistry, Peking University, Beijing 100871, China

Introduction

In *de novo* design of ion channel proteins, because of their rigid structures, tetrakis-phenylporphyrins have been employed to act as templates to combine α-helical peptide segments for the formation of bundle structures [1]. The aggregation state of the amphiphilic helix peptide can also be controlled by attaching four copies of it to this C_4-symmetric template. For this purpose, we synthesized a series of amino acid and oligopeptide-(p-,m-, o-,)substituted tetrakis-phenylporphyrin derivatives using different coupling agents (see Table 1).

Results and Discussion

1. According to the literature [2], compounds **1, 6** and **10** were prepared with the corresponding substituted benzaldehydes and pyrrole in 40-50% yields. The reduction of **1** and **6** by treating with $SnCl_2$ in concentrated hydrochloric acid obtained satisfactory yields (80-95%) of **2** and **7**. The compound **10** was then hydrolyzed with NaOH and pyridine to remove the methyl ester protection to get compound **11**. The products **2** and **7** were reacted with succinic anhydride to get 3 and **8**. The templates **3,7** and **11** were employed in peptide synthesis. Further condensation reactions with PheLeuOBzl, $GlyGlyOC_2H_5$, FmocGly and AlaOBzl got products **4, 5, 9, 12,14, 15** in good yield (65-86%).
2. Compounds **3, 7** and **11** were condensed with various amino acids or peptides by a new coupling agent 3-(Diethoxyphosphoryloxy)-1,2,3-benzotriazin-4(3H)-one (DEPBT), which was developed by our group [3]. Other compounds were prepared by DCC. All of these obtained compounds have been confirmed by MS, NMR, UV-Vis and some of them by elemental analysis.
3. Because of the poor solubility in most of the organic solvents, all of these substituted tetrakis-phenylporphyrin derivatives were difficult to purify. Also due to the incompleted substitutions at the four symmetric condensation positions of the porphyrins, we had to use silica columns or even HPLC to get the pure products.
4. Compounds **3** and **8** have four arms stretching out from four predetermined positions of tetrakis-phenylporphyrin templates. The attachments provide favorable interhelical spacing as well as a degree of conformational flexibility and more reaction activity. We supposed that they may be valuable starting units for membrane proteins containing a porphyrin ring. In the future, the synthesis of these compounds may find application in design and synthesis as the template for artificial proteins.

Table 1. The structures of amino acids, oligopeptide-tetrakis-phenylporphyrin derivatives

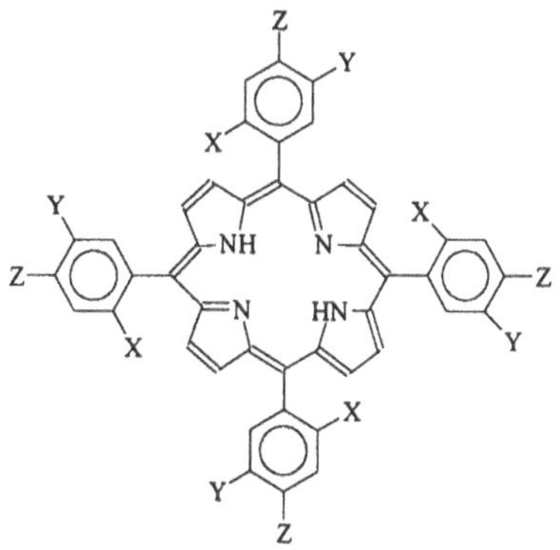

No.	X	Y	Z
1	NO_2	H	H
2	NH_2	H	H
3	$NHCOCH_2CH_2COOH$	H	H
4	$NHCOCH_2CH_2CO$PheLeuOBzl	H	H
5	$NHCOCH_2CH_2CO$GlyGlyOC_2H_5	H	H
6	H	NO_2	H
7	H	NH_2	H
8	H	$NHCOCH_2CH_2COOH$	H
9	H	NHGlyFmoc	H
10	H	H	$COOCH_3$
11	H	H	COOH
12	H	H	COOAlaOBzl
13	H	H	COOAlaOH
14	H	H	COOGlyGlyOC_2H_5
15	H	H	COOPheLeuOBzl

Acknowledgment

This research was supported by the National Key Projects for Fundamental Research.

References

1. Akerfeldt, K. S., Kim, R. M. Camac, D., Groves, J. T., Lear, J. D. and DeGrado, W. F., J. Am. Chem. Soc., 114 (1992) 9656.
2. Lindsey, J. S., Schreiman, I. C., Hsu, H. C., Kearney, P. C. and Marguerettaz, A., J. Org. Chem., 52 (1987) 827.
3. Fan, C.-X., Hao, X.-L. and Ye, Y.-H., Syn. Commun., 26(1996) 1455.

De novo design and synthesis of amphiphilic α-helical peptides

Xiao-Lin Hao, Xiao-Hua Chen and Yun-Hua Ye
Department of Chemistry, Peking University, Beijing 100871, China

Introduction

Previous studies indicate successful results on the bundles of amphiphilic transmembrane α-helices as a structural motif for ion-channel proteins. Thus, the *de novo* design and synthesis of amphiphilic α-helical peptides become practical ways for the artificial modeling of channel proteins [1]. On the basis of secondary structure prediction and molecular modeling, we design, synthesize and characterize three amphiphilic heneicosapeptides, LSSLLGLLSSLLP LLSSLLXX, X=K(I), X=E(II), EALAKLLEALAKLLEALA KLK(III).

Results and Discussion

1. These three 21-residue peptides were synthesized by means of stepwise solid-phase peptide synthesis. With the Boc protection strategy, all syntheses were carried out manually starting from Boc-protected amino acids covalently linked to Merrifield resin. A 3 or 4-fold excess of protected amino acid derivatives were coupled using DCC with HOBt as additive or 3-(diethoxyphosphoryloxy)-1,2,3-benzotriazin-4(3H)-one (DEPBT), a novel organophosphorus coupling reagent [2]. Coupling reactions were monitored by Kaiser test. Anhydrous HF was used to cleave the peptides from the resin. The crude products obtained were filtrated by Sephadex G-25 and further purified by RP-HPLC.
2. MS and amino acid analysis confirmed the integrity of these three amphiphilic peptides, the results are shown in Table 1. as follows:
The results of amino acid analysis:

Peptide (I): Ser 5.72(6), Gly 1.35(1), Pro 1.00(1), Leu 10.66(11), Lys 2.33(2)
Peptide (II): Ser 5.45(6), Gly 1.29(1), Pro 1.07(1), Leu 11.00(11), Glu 2.26(2)
Peptide (III): Glu 3.3(3), Lys 4.2(4), Ala 5.8(6), Leu 7.4(8)

3. Peptide (I) and (II) had poor solubility which resulted in great difficulties in purification. They had to be dissolved in 50% acetic acid and eluted by more than 20% acetic acid. In order to improve the solubility of Peptide (I) and (II) as well as enhance their helical degree, Peptide (III) was designed and synthesized by inserting Glu and Lys, two residues with oppositely charged side chains, in the sequence, with the expectation that the favorable intrachain charge-charge interactions would result in more solubility while favoring the formation of α-helix. Peptide(III) was readily purified by Sephadex G-25 and reversed-phase HPLC. It also showed stronger α-helical features than Peptide(I) and (II) by CD spectroscopy in both water and TFE (see Figure 1).

Table 1. The MS (FAB) of the three peptides

	Peptide(I)	Peptide(II)	Peptide(III)
M.W.	2194	2196	2247
MS(FAB)	$(M+H)^+$: 2195	$(M+H)^+$: 2197	$(M+Na+1)^+$: 2271

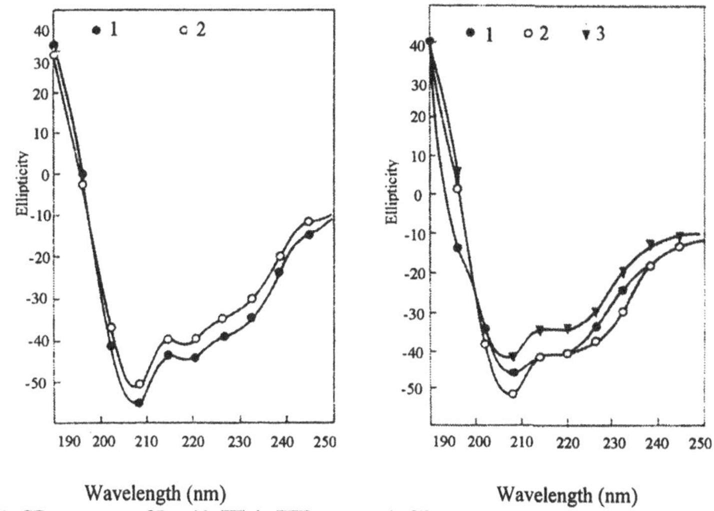

1. CD spectrum of Peptide(III) in TFE
2. spectrum of Peptide(III) in 20 mM Phosphate, 0,1 M kcl (pH=7.1)

1. CD spectrum of Peptide(II) (2000μg/ml) in TFE
2. CD spectrum of Peptide(III)(1000μg/ml) in TFE
3. CD spectrum of Peptide(I) (1000μg/ml) in TFE

Fig. 1. CD spectrum of each peptide in different conditions.

4. The abilities of the peptides to form stable amphiphilic α-helical bundles were characterized by circular dichroism studies in buffer or TFE solvents. Their CD curves displayed typical features of peptides in α-helical conformation, i.e., strong negative Cotton effects at 222 and 208nm, zero-crossover points at 202nm, and strong positive Cotton effects at 193 nm. And the formation of significant secondary structure in Peptide (III) both in TFE and in water reflects its helix-formation stability.

Acknowledgments

This research was supported by the National Key Projects for Fundamental Research. The authors thank Dr. Yu Luo and Dr. Chun-Xu Qu for instructive suggestion and helpful discussion on the structure design.

References

1. Oili, S., Madison, V. and Montal, M., Proteins: Structure, Function and Genetics, 8 (1990) 226.
2. Fan, C.-X., Hao, X.-L. and Ye, Y.-H., Syn. Commun., 26 (1996) 1455.

Session V
Bioactive Peptides

Chairs: John D. Wade
University of Melbourne
Victoria, Australia

Yoshiaki Kiso
Kyoto Pharmaceutical University
Kyoto, Japan

Gui-Shen Lu
Institute of Materia Medica
Beijing, China

and

Xiao-Jie Xu
Peking University
Beijing, China

Relaxin: Recent structure and activity studies

John D. Wade[a], Feng Lin[a], Yan-Yeow Tan[b], Gert Talbo[a] and Geoffrey W. Tregear[a]

[a]Howard Florey Institute of Experimental Physiology and Medicine [b]Department of Pharmacology, University of Melbourne, Parkville, Victoria 3052, Australia

Introduction

The peptide hormone relaxin is produced in, and acts upon, tissues of the reproductive tract in many mammalian species to facilitate parturition [1]. The primary structure of rat relaxin was determined by its isolation, sequencing and amino acid analysis [2]. It was shown to consist of a twenty four residue A-chain and a thirty five residue B-chain (Fig. 1). Comparatively little is known about the structure and biology of this peptide. It was shown to be active in both the mouse pubic ligament assay and the rat uterine contraction assay. However, it produced a non-parallel dose-response curve to the porcine relaxin standard [2]. Binding sites for relaxin have been detected in the uterus and, more recently, in the brain and heart of both male and female rats [3,4]. Subsequent work has shown that relaxin possesses potent inotropic and chronotropic activity in the isolated rat heart assay [5]. This suggests that it may, in fact, have a broad physiological role. Given the convenience of the rat as a laboratory model for biological study and, as part of a study of the role of rat relaxin in pregnancy, we undertook the solid phase synthesis of the peptide and compared it with the native product both chemically and biologically.

Methods

Each of the A- and B-chains were assembled by the Boc-polystyrene solid phase method as previously described [6]. The N-terminal residue of the A-chain was attached as the pyroglutamyl form. Following cleavage and deprotection of the chains by treatment with HF, each was S-reduced with dithiothreitol and then purified by preparative RP-HPLC. The chain combination was carried out in solution also as previously described [6]. The inotropic and chronotropic activities of the rat relaxin in the isolated rat heart were carried out as previously described [5]. All data were statistically analyzed by the Student's t-test.

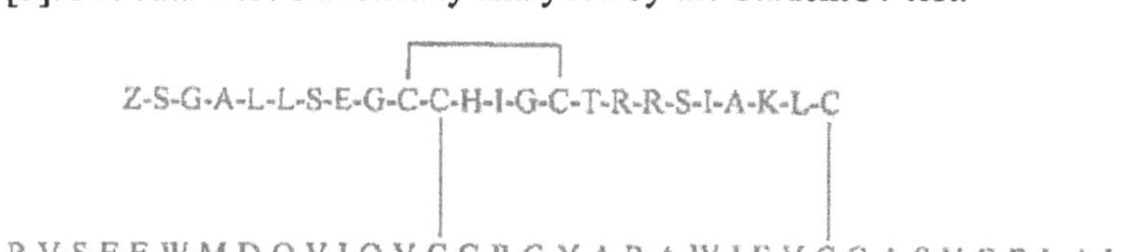

Fig. 1. Primary structure of rat relaxin.

Results and Discussion

The solid phase synthesis of rat relaxin proved to be difficult to successfully accomplish. The use of a two chain combination approach was again employed but the very poor solubility of the B-chain made both its purification and subsequent combination arduous with the result that the recovery was lower than previously achieved in our laboratory for other synthetic relaxins. Chain combination was less efficient than for other synthetic relaxins with the overall yield of final purified product being approximately 0.5%. The synthetic rat relaxin was characterized by a variety of analytical criteria including by MALDI-MS which gave a measured value of 6461.1 (theory: 6460.5). The synthetic peptide was also indistinguishable from the native peptide including by limited tryptic digestion followed by mass spectrometric analysis of the digests and by circular dichroism spectroscopy (data not shown).

Recently, the successful total chemical synthesis of human insulin was achieved via an elegant three-step regioselective disulfide bond approach [7]. However, attempts to adapt this method to the synthesis of rat relaxin were not successful, again because of the poor solubility of the selectively S-protected B-chain. More recent efforts to overcome this limitation have focused on the use of the novel secondary amide protecting group, the N-(2-hydroxy-4-methoxybenzyl) (Hmb) [8]. The Hmb function is compatible with the Fmoc/t-butyl method of solid phase peptide synthesis and inhibits chain aggregation during peptide assembly. Acetylation of the group makes it resistant to final peptide-resin cleavage and deprotection and the resulting Ac-Hmb-containing peptide has substantially improved solubility in aqueous solution.

Both native and synthetic rat relaxins were equally active in the *in vitro* rat heart assay. The two peptides were highly inotropic and chronotropic but less so than either the synthetic human H2 or H1 relaxins (Table 1). The pD2 values are higher than for endothelin or angiotensin II. The efficacy of the relaxins was excellent with all compounds producing 60-80% of the maximum response caused by the β-adrenoceptor agonist isoprenaline. Each relaxin had higher efficacy in the chronotropic than inotropic assay. The potent effect on the heart is probably by stimulation of a specific relaxin receptor for these are not abolished by the structurally similar insulin nor by the individual relaxin chains [9]. More recent work in our laboratory has shown

Table 1. Comparison of the inotropic and chronotropic effects of synthetic (s) and native (n) rat relaxins. Other peptides: human Gene 2 and human Gene 1 relaxins.

	Maximal response		pD2		Efficacy(isoprenaline=1)	
	inotropic	chronotropic	inotropic	chronopic	inotropic	chronotroic
hRlx1(B29)	0.21±0.03	130±7	9.00±0.08	9.05±0.10	0.70	0.77
hRlx2(B28)	0.19±0.02	100±10	8.81±0.04	8.73±0.05	0.45	0.60
n rat Rlx	0.24±0.06	117±17	8.59±0.03	8.51±0.07	0.55	0.69
s rat Rlx	0.24±0.02	123±13	8.72±0.03	8.60±0.02	0.75	0.78

that synthetic rat relaxin, when introduced into the lateral ventricle of the rat brain, causes significant production of *fos* in certain discrete regions known to regulate fluid balance and cardiovascular control. This implies that it has a direct central role, the significance of which requires further study [10]. The availability of the peptide will help shed further light on the biological role of the hormone.

Acknowledgments

We thank Mick Petrovski and Daniela Salvatore for excellent technical assistance, and gratefully acknowledge the provision of the CD spectra by Professor Laszlo Otvos, Jr. (The Wistar Institute, Philadelphia, USA). The work was supported by an NHMRC Institute Block grant.

References

1. Sherwood, O.D., In: Knobil, E. and Neill, J.D., (Eds.) The Physiology of Reproduction, 2nd ed., Raven
2. Press, New York, USA,1994, p. 861.
3. Walsh, J.R. and Niall, H.D., Endocrinol. 102 (1980) 1258.
4. Osheroff, P. and Phillips, H.S., Proc. Natl. Acad. Sci. USA, 88 (1991) 6413.
5. Osheroff, P.L., Cronin, M.J. and Lofgren, J.A., Proc. Natl. Acad. Sci. USA, 89 (1992) 2384.
6. Kakouris, H., Eddie, L.W. and Summers, R.J., Lancet 339 (1992) 1076.
7. Wade, J.D., Lin, F., Salvatore, D., Otvos, L.Jr., and Tregear, G.W., Biomed. Pept. Prot. Acids in press.
8. Akaji, K., Fujino, K., Tatsumi, T. And Kiso, Y., J. Am. Chem. Soc., 115 (1993) 11384.
9. Johnson, T., Quibell, M. and Sheppard, R.C., J. Pept. Sci., 1 (1995) 11.
10. Summers, R.J., Tan, Y.Y., Kakouris, H. and Eddie, L.W. In: MacLennan, A.H., Tregear, G.W. and Bryant-Greenwood, G.D. (Eds.), Progress in Relaxin Research. Global Publication Services, Singapore, 1995, 487.
11. McKinley, M.J., Burns, P., Colvill, L., Oldfield, B.J., Weisenger, R.S., Wade, J.D. and Tregear, G.W.1996, Brain Res. (submitted).

Novel synthesis of cyclotheonamide B1

Jin-Gen Deng[a], Yasumasa Hamada[b] and Takayuki Shioiri[b]
[a]*Chengdu Institute of Organic Chemistry, Academia Sinica, Chengdu 610041, China*
[b]*Faculty of Pharmaceutical Sciences, Nagoya City University, Tanabe-dori, Mizuho-ku, Nagoya 467, Japan*

Introduction

Cyclotheonamides (**1**, A: R=CHO; B: R=Ac) are macrocyclic thrombin inhibitors having two novel amino acid residues (Fig. 1). Their structural uniqueness and inhibiting activities have prompted us [1,2] and other groups to study these compounds [3]. In the synthesis of arginine-containing peptides, ornithine formation is often encountered. The possibility of preparing arginine-containing peptides by guanidination of the δ-amino groups of the appropriate ornithine-containing precursors has long been recognized. Such a strategy is attractive for its potential to eliminate the many problems associated with the use of conventionally protected arginine starting materials in peptide synthesis [4]. We report here a new approach for the synthesis of cyclotheonamide B (**1**, R=Ac) via the guanidination of the ornithine-containing macrocyclic peptide (**2**).

Results and Discussion

Deprotection of the dipeptide (**3**)[2] at the N^{β}-position and the tripeptide (**4**)[1a] at the C-terminal and subsequent coupling with DEPC gave the linear pentapeptide (**5**). The best result (71% yield) was obtained by using trimethylsilyltriflate for the deprotection of Boc group in comparison with the use of trifluoroacetic acid (48%) or *p*-toluenesulfonic acid monohydrate (43%). After the deprotection of pentapeptide (**5**) at the C- and N-terminal, the cyclization of (**5**) was proceeded by using FDPP to give the best yield (57%) and the

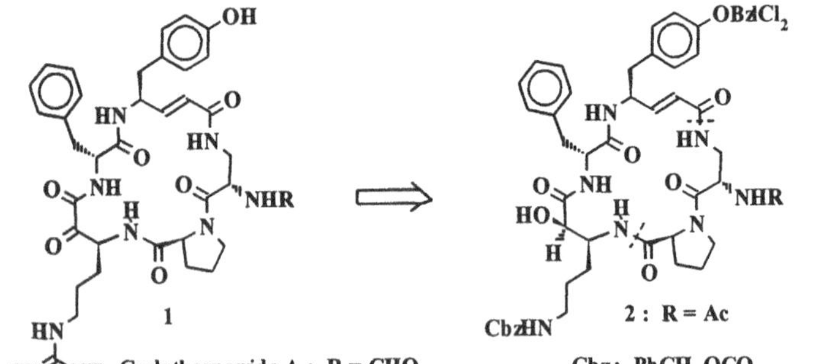

1

Cyclotheonamide A : R = CHO
Cyclotheonamide B : R = Ac

Cbz: $PhCH_2OCO$
$BzlCl_2$: 2,6-dichlorobenzyl

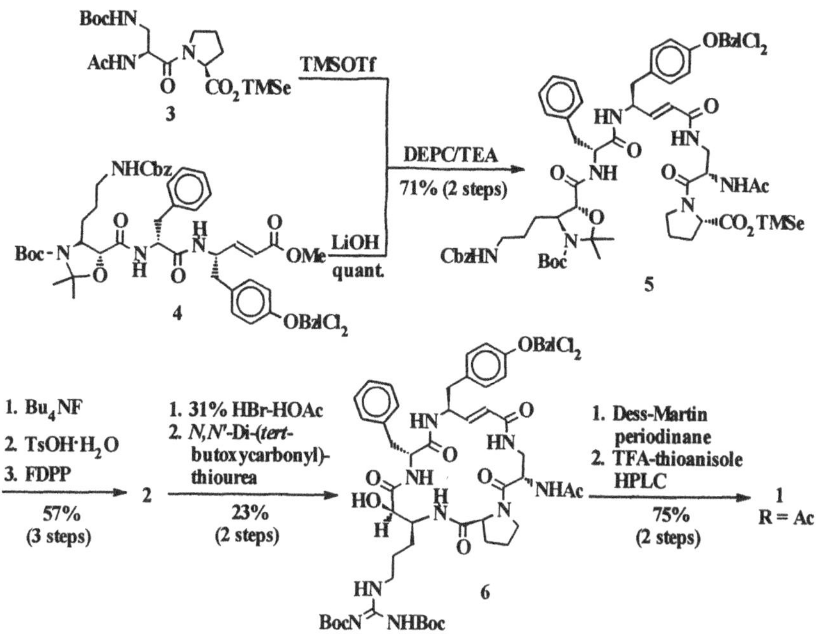

cyclization seems to be enhanced by the appropriate conformation of the linear peptide in solution and the efficiency of coupling reagents [1a, 5]. The cyclic peptide (**2**) was treated with 31% Hbr-HOAc and subsequent guanidination with *N,N'*-di-(*tert*-butoxycarbonyl) thiourea to give the arginine-containing macrocyclic peptide (**6**). The selective deprotection of Cbz amido group and benzyl-type ether group was also achieved by catalytic hydrogenation in the presence of 1,4-cyclohexadiene, and unfortunately the partial reduction of the substituent chlorine on $BzlCl_2$ group resulted in side reactions. Finally, oxidation of the α-hydroxy group of (**6**) with Dess-Martin periodinane, followed by *O, N*-deprotection, produced cyclotheonamide B (**1**, R=Ac) in 75% yield.

References

1. a)Deng, J., Hamada, Y. and Shioiri, T., Tetrahedron Lett., 37 (1996) 2261.
2. b)Bastiaans, H.M.M, van der Baan, J.L. and Ottenheijm, H.C.J., Tetrahedron Lett., 36 (1995) 5963.
3. Deng, J., Hamada, Y., Shioire, T., Matsunaga, S and Fusetani, N., Angew. Chem Int. Ed. Engl., 33 (1994) 1729.
4. Wipf, P., Chem. Rev., 95 (1995) 2115.
5. Bernatowicz, M.S., Wu, Y., and Matsueda, G.R., J., Org. Chem., 57 (1992) 2497.
6. Ehrlich, A., Rothemund, S., Brudel, M., Beyermann, M., Carpino, L. A. and Bienert, M., Tetrahedron Lett., 34 (1993) 4781.

Identification of superantigen TSST-1 sequence important for induction of T cell proliferation using synthetic peptides and antibody to CD28

Wei-Gang Hu and Xi-Hua Zhu

Department of Immunology, Third Military Medical University, Chongqing 630038, China

Introduction

Superantigens (SAg) are new protein antigens [1]. A body of evidence has implicated the potential role of SAg in various human diseases [2]. However, little has been known so far about the mechanism of molecular recognition between T cells and SAg. Toxic shock syndrome toxin-1 (TSST-1) belongs to the group of SAg. It is a 22 kDa single-chain polypeptide containing 194 amino acid residues and can activate human T cells bearing $V_{\beta}2$. In this study, we used a new method employing synthetic peptides and antibody (Ab) to the CD28 to identify TSST-1 sequence involved in its binding to the T cell receptor.

Results and Discussion

It is feasible to use the synthetic peptide technique to locate T cell epitopes of protein antigens (Ag), but difficult to locate T cell epitopes of SAg [3]. T cell activation by Ag requires the assistance of antigen presenting cells (APC) that express MHC molecules. Thus, Ag must possess both T cell epitopes and MHC binding sites. These two sites in classical Ag are overlapping within a short peptide (10 residues or so) [4], whereas those in SAg are away from each other [5]. Therefore, it is difficult for a synthetic peptide of limited length to contain the two kinds of sites of SAg. Recently we and others have demonstrated that APCs assist T cell activation by SAg with CD28 costimulation and the cross-linked Ab to CD28 can replace APC to assist T cell activation by SAg [6]. For these reasons, we added Ab to CD28 to the synthetic peptide to establish a new method of locating T cell epitopes for SAg.

TSST-1(125-158) and TSST-1(101-158) could not activate peripheral blood mononuclear cells (PBMC), but could do so with CD28 costimulation, indicating that TSST-1(125-158) and TSST-1(101-158) contain T cell epitopes but not MHC binding sites. The sequence of TSST-1(125-158) was included in TSST-1(101-158). With CD28 costimulation the former has much better ability to activate T cells than the latter. It is possible that the conformation of TSST-1(101-158) is unfavourable for its T cell epitope to bind T cells. On the other hand, TSST-1(125-158) and TSST-1(101-158) plus CD28 costimulation could not activate $V_{\beta}2$ T cell-depleted PBMC, that is, the specificity of TSST-1(125-158) and TSST-1(101-158) to V_{β} element is identical to native TSST-1 molecule.

Our data suggest that TSST-1(125-158) contains SAg T cell epitope of and that the privide method using synthetic peptides and Ab to CD28 is effective to locate T cell

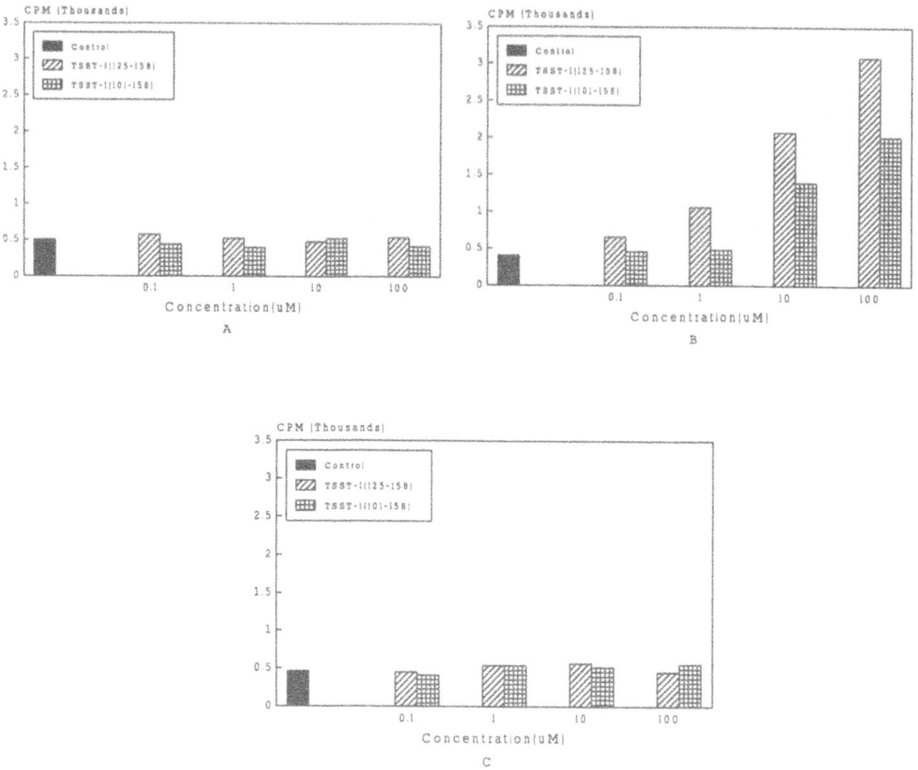

Fig. 1. Effects of synthetic peptides on T cell proliferation. Proliferation of PBMC was quantitated by incorporation of radiolabel [counts per minute(CPM)] into cellular DNA. (A) Stimulation of PBMC by peptides alone. (B) Stimulation of PBMC by peptides plus CD28 costimulation. (C) Stimulation of $V_{\beta}2$ T cell-depleted PBMC by peptides plus CD28 costimulation.

epitopes of SAg. Our studies provide a foundation for further studying SAg T cell epitopes of TSST-1.

Acknowledgments

We thank Dr.J.A.Ledbetter for his gift of Ab to CD28 and Dr.F.C.Lancaster for Ab to $V_{\beta}2$.

References

1. Marrack, P. and Kappler, J., Science, 248 (1990)705.
2. Acta-Orbea, H., Ann. Rheum. Dis., 52 (1993) S6.
3. Ramesh, N., Parronch, F.P., Ahern, D., Romagnani, S. and Geha, R., Immunol.Invest., 23(1994)381.
4. Rothbard, J.B. and Gefter, M.L., Annu. Rev. Immunol., 9 (1991) 527.
5. Acharya, K.R., Passalacqua, E.F., Jones, E.Y., Harlos, K., Stuart, D.I., Brehm, R.D. and Tranter, H.S., Nature, 367 (1994) 94.
6. Ohnishi, H., Ledbetter, J.A., Kanner, S.B., Linsley, P.S., Tanaka, T., Geller, A.M. and Kotb, M., J. Immunol., 154 (1995) 3180.

Analogs of the δ opioid antagonist TIPP with increased lipophilic character

P. W. Schiller, G. Weltrowska, I. Berezowska, T. M.-D. Nguyen, N. N. Chung, C. Lemieux, K. A. Carpenter and B. C. Wilkes
Laboratory of Chemical Biology and Peptide Research, Clinical Research Institute of Montreal, 110 Pine Avenue West, Montreal (Que.), Canada H2W 1R7

Introduction

δ-Opioid antagonists are of interest not only as pharmacological tools but also as potential therapeutic agents. Most importantly, chronic co-administration of a δ antagonist and morphine was shown to greatly attenuate the development of morphine tolerance and dependence [1,2]. Furthermore, antagonists may be useful for the treatment of cocaine and alcohol addiction and as immunosuppressants in organ transplantation. Among the currently available antagonists, the enkephalin analog ICI 174864 is quite δ-selective but not very potent, whereas the naltrexone derivative naltrindole is highly potent but not very δ-selective. Recently, the tetrapeptide H-Tyr-Tic-Phe-Phe-OH (TIPP; Tic = tetrahydroiso-quinoline-3-carboxylic acid) was found to be a potent and highly selective δ antagonist [3]. The pseudopeptide H-Tyr-TicΨ [CH_2-NH]Phe-Phe-OH (TIPP[Ψ]) showed the same high δ antagonist potency as TIPP, unprecedented δ selectivity and high stability against chemical and enzymatic degradation [4].

Even though TIPP and TIPP[Ψ] are relatively lipophilic peptides, their ability to cross the blood-brain barrier (BBB) may still be somewhat limited. In the case of TIPP[Ψ] the secondary amino group of the reduced peptide bond is protonated at physiological pH and, therefore, this pseudopeptide is somewhat more polar than the TIPP parent peptide. In an effort to enhance the lipophilicity of TIPP and TIPP[Ψ], and their ability to penetrate the BBB, we prepared analogs that contain various alkyl-, aralkyl- or halogen substituents. Alkyl- or aralkyl groups were introduced at the N-terminal amino group of TIPP or at the secondary amino group of TIPP[Ψ]. Systematic halogenation with F, Cl, Br or I in *para*-position of the aromatic ring of either Phe^3 or Phe^4 of TIPP was also performed.

Results and Discussion

The peptides and pseudopeptides were synthesized by the solid-phase method according to published protocols [3,4]. The N-terminally substituted TIPP analogs were prepared by reacting the N-terminal amino group of the resin-bound tetrapeptide with the appropriate aldehyde, followed by reductive alkylation with sodium borohydride [5]. The pseudopeptides containing a reduced and *N*-alkylated peptide bond (Ψ [CH_2-NR]) in the 2-3 position of the peptide sequence were synthesized by first performing a reductive alkylation reaction between 2-Boc-1,2,3,4-tetrahydroisoquinoline-3-aldehyde and the amino group of the resin-bound H-Phe-Phe dipeptide[4], followed by a reductive alkylation reaction between the secondary amino group of the reduced peptide bond and the appropriate aldehyde [6], and

Table 1. In vitro opioid activities and hydrophobicity parameters of TIPP analogs

	MVD assay	Opioid receptor binding[c]			HPLC
Compound	Ke [nM][a]	Ki^{μ}[nM]	Ki^{δ}[nM]	Ki^{μ}/Ki^{δ}	k'[e]
Tyr(N(CH$_2$)$_5$CH$_3$)-Tic-Phe-Phe-OH	4.28	1080	1.10	982	5.25
Tyr(NCH$_2$-cyclopropyl)-Tic-Phe-Phe-OH	28.2[b]	6910	4.84	1430	5.79
Tyr(N(CH$_2$CH$_3$)$_2$-Tic-Phe-Phe-OH	3.39	7940	1.22	6510	6.22
H-Tyr-TicΨ[CH$_2$-NCH$_3$]Phe-Phe-OH[d]	2.89[b]	13400	0.842	15900	4.46
H-Tyr-TicΨ[CH$_2$-NCH$_2$CH$_3$]Phe-Phe-OH	14.4	4530	6.31	718	4.75
H-Tyr-TicΨ[CH$_2$-N(CH$_2$)$_5$CH$_3$]Phe-Phe-OH	13.1	18800	66.6	282	7.57
H-Tyr-TicΨ[CH$_2$-NCH$_2$CH$_2$Ph]Phe-Phe-OH	13.8[b]	1130	4.51	251	7.89
H-Tyr-Tic-Phe(*p*-F)-Phe-OH	1.62	5480	1.65	3320	5.44
H-Tyr-Tic-Phe(*p*-Cl)-Phe-OH	1.60	8130	1.26	6450	5.70
H-Tyr-Tic-Phe(*p*-Br)-Phe-OH	3.35	1200	0.382	3140	5.86
H-Tyr-Tic-Phe(*p*-I)-Phe-OH	2.88	1190	0.570	2090	6.21
H-Tyr-Tic-Phe-Phe(*p*-F)-OH	2.97	3210	1.26	2550	5.70
H-Tyr-Tic-Phe-Phe(*p*-Cl)-OH	2.52	2070	1.02	2030	6.48
H-Tyr-Tic-Phe-Phe(*p*-Br)-OH	2.93	2870	1.42	2020	6.69
H-Tyr-Tic-Phe-Phe(*p*-I)-OH	2.37	2690	0.509	5280	6.85
H-Tyr-Tic-Phe-Phe-OH (TIPP)	4.80	1720	1.22	1410	5.14
H-Tyr-TicΨ[CH$_2$-NH]Phe-Phe-OH (TIPP[Ψ])	2.89	3230	0.308	10500	4.11

[a]Determined against DPDPE. [b]Determined against [D-Ala^2]deltorphin I. [c]Displacement of [^{3}H]DAMGO (μ-selective) and [^{3}H]DSLET (δ-selective) from rat brain membrane binding sites. [d]Ref 7. [e]Vydac column 218TP54 (4.6 x 250 mm); I, gradient 20-50% B in 20 min (A - 0.1% TFA in H$_2$O, B - 0.1% TFA in acetonitrile); II, isocratic 50% A/50%B; flow rate 1 mL/min; detection 215 nm.

subsequent addition of the N-terminal tyrosine residue. Peptides were cleaved from the resin by HF/anisole treatment in the usual manner. Crude products were purified by reversed-phase chromatography.

Substitution of an *n*-hexyl group or of two ethyl groups at the N-terminal amino function of TIPP resulted in compounds which retained antagonist potency comparable to that of the TIPP parent peptide, as determined with the mouse vas deferens (MVD) assay. The analog carrying a cyclopropylmethyl group at the N-terminus was a somewhat less potent δ antagonist. In the receptor binding assays all three N-terminally substituted TIPP analogs showed high δ receptor affinities and excellent δ receptor selectivities (Ki^{μ}/Ki^{δ} = 1000-6500) (Table 1). Introduction of a methyl substituent at the secondary amino group of the reduced peptide bond of TIPP[Ψ] had previously been shown to produce a compound with unchanged δ antagonist potency, high δ receptor affinity and unprecedented δ receptor selectivity [7]. Analogous substitution of TIPP[Ψ]with an ethyl-, *n*-hexyl- or phenethyl group led to compounds that showed only slightly reduced δ antagonist potency as compared to TIPP or TIPP[Ψ]. The *N*-ethyl and *N*-phenethyl-substituted TIPP[Ψ]analogs retained high

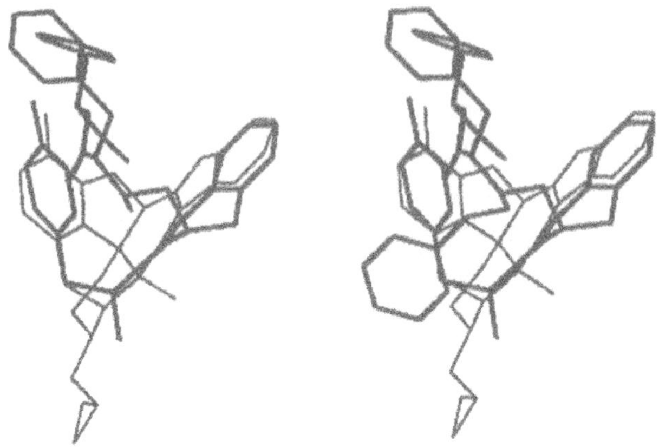

Fig. 1. Superimposition of naltrindole (drawn in light lines) with low energy conformers of H-Tyr-Tic Ψ[CH_2-N^+HCH_3]Phe-Phe-OH (left panel) and H-Tyr-Tic Ψ[CH_2-$N^+HCH_2CH_2Ph$]Phe-Phe-OH (right panel). Peptide structures are drawn in heavy lines.

δ receptor affinity, whereas the *N*-*n*-hexyl derivative bound somewhat less tightly to the δ receptor. These substituted pseudopeptides all had very weak affinity for μ receptors and, therefore, still showed high receptor selectivity. Systematic halogenation with F, Cl, Br or I in *para*-position of the aromatic ring of either Phe^3 or Phe^4 of TIPP resulted in eight compounds that all showed slightly higher δ antagonist potency than the parent peptide. In comparison with TIPP, these halogenated peptides displayed either the same or slightly higher δ receptor affinity and higher δ receptor selectivity in all eight cases.

Determination of the hydrophobicity parameters k' by HPLC revealed that the performed N-terminal substitutions in TIPP resulted in compounds with markedly enhanced lipophilic character (Table 1). A very drastic increase in lipophilicity was observed upon introduction of an *n*-hexyl or phenethyl substituent at the secondary amino group of TIPP[Ψ]. Substitution of F, Cl, Br or I in *para*-position of the aromatic ring of Phe^3 or Phe^4 of TIPP also produced a progressive increase in lipophilic character in both cases, whereby the Phe^4 halogenated analogs showed a somewhat more pronounced lipophilicity enhancement than the corresponding Phe^3 halogenated analogs.

At physiological pH the tertiary amino group of the peptides containing an *N*-alkylated reduced peptide bond may exist as two interconverting diastereomers (R and S configuration at the nitrogen). Molecular mechanics and molecular dynamics studies were performed on both diastereomers of H-Tyr-TicΨ[CH_2-N^+HCH_3]Phe-Phe-OH and H-Tyr-TicΨ[CH_2N^+H CH_2CH_2Ph]Phe-Phe-OH, and in both cases resulted in low energy conformers that showed good spatial overlap between the Tyr^1 and Tic^2 aromatic rings and N-terminal amino group of the pseudopeptides with the corresponding aromatic rings and nitrogen atom of the non-peptide δ antagonist naltrindole (Fig. 1). These low energy conformers closely resemble a previously proposed model of the δ receptor-bound conformation of the TIPP[Ψ] parent peptide [8]. This finding provides an explanation for the fact that *N*-alkylation of the reduced peptide bond in TIPP[Ψ] results in compounds that retain high δ antagonist potency and δ receptor selectivity. The results of molecular dynamics studies performed with TIPP[Ψ] analogs containing an *N*-alkylated reduced peptide bond in protonated form (R and S configuration) showed that the "ω_2" torsional angle around the $-CH_2-N^+HR-$ bond predominantly assumed a value of -150° in the case of the R configuration and a value of +150° in the case of the S configuration. These values of "ω_2" are close to the 2 value of 180° in a regular *trans* peptide bond. Fluctuations to "ω_2" values of -60° (R configuration) and +60° (S configuration) were seen along the dynamics trajectory. These fluctuations were relatively less frequent than in the case of TIPP[Ψ] in the protonated state, indicating that N-alkylated reduced peptide bonds are structurally somewhat less flexible than the non-alkylated reduced peptide bond.

In summary, the performed introduction of alkyl-, aralkyl or halogen substituents at various positions of the TIPP or TIPP[Ψ] molecules resulted in compounds that showed much increased lipophilic character and retained high δ antagonist potency and δ receptor selectivity. These various substitutions by themselves or in combination can be expected to improve penetration of the BBB via passive diffusion.

Acknowledgment

This work was supported by grants from the MRCC (MT-10131) and NIDA (DA-04443).

References

1. Abdelhamid, E.E., Sultana, M., Portoghese, P.S. and Takemori, A.E., J. Pharmacol. Exp. Ther., 258 (1991) 299.
2. Fundytus, M.E., Schiller, P.W., Shapiro, M., Weltrowska, G. and Coderre, T., Eur. J. Pharmacol., 286 (1995) 105.
3. Schiller, P.W., Nguyen, T.M.-D., Weltrowska, G., Wilkes, B.C., Marsden, B.J., Lemieux, C. and Chung, N.N., Proc. Natl. Acad. Sci. USA, 89 (1992) 11871.
4. Schiller, P.W., Weltrowska, G., Nguyen, T.M.-D., Wilkes, B.C., Chung, N.N. and Lemieux, C., J. Med. Chem., 36 (1993) 3182.
5. Summers, M.C. and Hayes, R.J.,J. Biol. Chem., 256 (1981) 4951.
6. Harbeson, S.L., Shatzer, S.A., Le, T.-B. and Buck, S.H., J. Med. Chem., 35 (1992) 3949.
7. Schiller, P.W., Schmidt, R., Wilkes, B.C., Weltrowska, G., Nguyen, T.M.-D., Chung, N.N., and Lemieux, C. In Lu, G.-S., Tam, J.P. and Du, Y.-C. (Eds.) Peptides: Biology and Chemistry (Proceedings of the 1994 Chinese Peptide Symposium), ESCOM, Leiden, The Netherlands, 1995, p. 140.
8. Wilkes, B.C. and Schiller, P.W., Biopolymers (Peptide Science), 37 (1995) 391.

Cetrorelix, a potent LHRH-antagonist: Chemistry, pharmacology and clinical data

J. Engel and Th. Reissmann
Corporate Research & Development ASTA Medica AG, Dresden, Germany

Introduction

The peptide hormone LHRH (LH releasing hormone) is liberated by hypothalamic neurons and regulates the release of the gonadotropins luteinizing hormone (LH) and follicle stimulating hormone (FSH) from the anterior pituitary gland. In 1971 Schally et al. [1] were able to isolate LHRH from hypothalamic extracts and to establish its amino acid sequence. The replacement of different amino acids then led to the discovery of highly active analogues like the superagonists, which are used clinically today in many hormone-dependent diseases. Paradoxically, by chronic administration these LHRH-agonists can also be used to suppress the gonadotropins, a mechanism which is based on the induction of a down-regulation of pituitary receptors. However, the administration of agonists initially leads to a stimulatory effect on LH, FSH and sex steroid secretion, which might result in a "flare up" of the disease. The advantage of antagonistic analogues is the immediate onset of gonadal steroid suppression by competitive binding to the LH-RH receptors already after a single administration. So far the clinical development of LHRH-antagonists was delayed due to low potency, side effects such as edema due to histamine release and high dosage requirements [2]. With the synthesis of compounds like Nal-Glu or Antide these side-effects started to decrease. Major improvement was achieved by the new LHRH-antagonist Cetrorelix (INN). This antagonist is free of anaphylactoid effects and produced a strong suppression of gonadotropins [3]. This paper summarises results from the chemical, pharmacological and clinical development of Cetrorelix.

Results and Discussion

Synthesis and stability

As the natural LHRH, Cetrorelix is a decapeptide but within the molecule 5 amino acids were replaced by unnatural D-amino acids [(Ac-D-Nal(2)^1-D-Phe(4Cl)^2-D-Pal(3)^3-D-Cit^6-D-Ala^{10})-LH-RH]. The peptide has a molecular weight of 1431.07 and was synthesised both, in classical and solid-phase pathways. The Merrifield synthesis was performed with methylbenzhydrylamino-polystyrene-divinylbenzene resin (MBHA resin), using the usual methods of sequential synthesis for oligopeptides. Boc-protection was cleaved quantitatively by 50 % trifluoroacetic acid in CH_2Cl_2, followed by neutralization with 10 % diisopropylethylamine (DIPEA) in CH_2Cl_2. The crude peptide was resolved in aqueous acetic acid for separation from the resin, precipitated *in vacuo* and purified via preparative HPLC using standard reverse phase routines.

The chemical and physical stability of Cetrorelix has been characterized in different test systems. Aqueous stability of Cetrorelix was investigated in the range of pH 1.0 to 13.0. The peptide was found to be surprisingly stable at pH 7.0 for a period of 21 days and is resistant to oxidative attack with H_2O_2 under neutral conditions. Peptides and proteins in general undergo proteolysis by enzymes, with severe consequences for the bioavailability of the drugs. Cetrorelix is highly resistant to degrading enzymes such as Chymotrypsin, Pronase and Nargase (Subtilisin) for several days which is in sharp contrast to LH-RH agonists facing almost complete degradation within several hours.

Pharmacology

Cetrorelix showed a high binding affinity to rat pituitary receptors and thus, a pronounced and immediate suppressive effect on the gonadotropins (LH, FSH) and subsequently on sex-hormones was observed in rodent animal models and in monkeys, in which Cetrorelix was most effective as compared to other antagonists. The strong suppression of sex-steroids resulted in a complete tumor regression in the DMBA-induced mammary carcinoma of rats and similar growth inhibitory effects were found in the MXT mammary carcinoma of mice, the Dunning prostate carcinoma of rats and on different human ovarian cancer cell lines. In DMBA-tumor bearing rats much lower doses proved to be sufficient and a prolonged duration of action was observed as compared to other antagonists. When single high doses of 1.0 mg/kg, 3.16 mg/kg and 10.0 mg/kg were injected subcutaneously, a dose- dependent duration of growth inhibition and tumor regression was found, lasting for 43 days in the highest dose group. The LH-RH antagonist Antide, which was tested in comparison, was less effective with regard to both parameters, tumor regression and duration of tumor growth inhibition. The antitumor response obtained with a single dose of 1 mg/kg Cetrorelix could be maintained and even became more pronounced, when this dose was given once a week for 16 weeks without any sign of escape. We then examined the efficacy of dose-fractionation. When the treatment was started with a single loading dose of 3.16 mg/kg, surprisingly, a daily dose of 31.6 μg/kg, equivalent to 1/100 of the initial dose and given from day 7 to 21, was effective in maintaining tumor regression. Microscopically the effects of Cetrorelix on this tumor were characterized by a loss of mitotic activity, marked atrophy with apoptosis and normalization and differentiation towards a normal mammary architecture. Interestingly, a significant antitumor activity was also found in ovarian cancer xenografts and hormone independent animal tumor models, as the Dunning prostate cancer of rats. In both tumor models a decrease in tumor weight and a prolongation in survival time of the animals was found. For both tumor models direct cytostatic/cytotoxic effects can be assumed, since the measurement of EGF revealed a loss of the corresponding receptor as well as a decrease in EGF-levels in the tumor tissues after treatment with Cetrorelix. Thus, an interference of the antagonist with growth factors might contribute to the observed effects. In contrast to former antagonists, Cetrorelix showed a very low potency to induce a histamine release from mast cells in vitro and subsequently, no anaphylactoid reactions were seen *in vivo*. In safety pharmacological as well as in 6 month toxicological studies in rats and dogs, Cetrorelix exerted no systemic side effects. In addition, it was shown to be non-teratogenic and no mutagenic potential was observed. Those effects which are related to the pharmacological action of the compound (reduction in weight of testis, ovaries, uterus, sperm counts) were reversible, with the time period needed for recovery being dependent on the treatment duration. Based on this favourable safety-profile, clinical Phase-I studies were initiated in volunteers. In these studies a lyophilisate of Cetrorelix-acetate was used.

Summary of clinical results

The safety of Cetrorelix was also confirmed in clinical Phase-I trials in which more than 100 volunteers, male and female, were treated with Cetrorelix. Even in supraoptimal doses, the tolerability proved to be very good. The dose range tested for single doses in males was 0.25 mg up to 20 mg. Compared to the placebo group, the extent and duration of suppression showed to be dose-dependent. After the administration of 1.0 to 5 mg Cetrorelix sc., maximal testosterone suppression was seen 8 to 12 hours after the injection. Twenty-four hours after administration of 1.0 and 2.0 mg Cetrorelix, testosterone values were no longer different from those in the placebo group, whereas in the 5.0-mg dose group testosterone concentrations increased slightly and reached serum concentrations within the lower normal range after 48 h. A linear kinetic relationship was found, with a calculated plasma t½ of 30h after single doses of 5 mg. During daily doses for 8 days, again a dose-dependent suppression was found but interestingly only a dosage as high as 10 mg/d was able to maintain testosterone levels within the castration range over the time period of administration [4]. With the use of lower doses an increase of testosterone levels was found between days 2-4 of treatment. Therefore, based on results from animal studies and pharmacokinetic considerations, it was tested whether a suppression of LH, FSH, and testosterone can be achieved by initial high-dose Cetrorelix and maintained by continued low-dose injections. Subsequently, a loading-dose schedule applying the 10 mg/d dose for 5 days followed by a maintenance dose of 1 mg/d was then tested in male volunteers and resulted in a continuous suppression of testosterone [5]. In addition, in these studies it was found that treatment with Cetrorelix induced a strong (appr. 40%), reversible reduction of the prostate volume within 2 weeks of treatment. Whether a receptor down regulation induced by Cetrorelix is the basis for the efficacy of low maintenance dosages is subject to further investigations. Phase-I studies performed in Europe and Japan revealed that this loading-dose treatment was not necessary in women. In female volunteers a suppression of estradiol for about one week was achieved after single sc. doses of 3 to 5 mg.

In Phase-II trials a beneficial effect of Cetrorelix was also found in patients suffering from benign prostate hyperplasia and prostate cancer [6]. In BPH patients, the efficacy with regard to improvement in symptoms was independent of the degree of testosterone suppression and a short term treatment of 4 weeks already resulted in an immediate relief of clinical symptoms lasting for several months in some cases. In addition, clinical studies were carried out in women suffering from tubal infertility who are treated with hormones in order to induce multiple follicular maturation for assisted reproduction (COS/ART). The classical stimulation procedure with gonadotropins (HMG, HCG) has the disadvantage of an unpredictable ovarian reaction and the occurrence of premature LH-surges. The endogenous LH-search can result in luteinization with a negative impact on the quality of the oocytes, which in turn result in the cancellation of the treatment cycle. Presently, the so called long protocol of LHRH-agonists, in which the treatment with agonists starts at least 14 days before the stimulation with gonadotropins, proved to be the most effective procedure. However, disadvantages of this treatment schedule are the long treatment period and the high doses of gonadotropins necessary. The use of Cetrorelix for the avoidance of premature LH-surges in in-vitro fertilisation programs showed some substantial improvements. By a short term treatment LH-surges could be avoided in about 200 patients by using daily doses as low as 0.25 mg or a single dose of 3 mg. The pregnancy rate so far, is comparable with standard treatment and there is a chance to reduce overall treatment costs by the possible reduction in the amount of HMG needed. The unwanted stimulatory phase of the LHRH-agonists can be avoided and the prolonged treatment duration can be reduced significantly [7-9]. Phase-III studies in this indication are ongoing.

Potential depot formulation of Cetrorelix

The LHRH-agonists currently on the market are applied as nasal sprays, sc.-implants or im. depot formulations based on poly-lactide / poly-glycolide matrices. The monthly dose of these formulations is 3.8 mg, whereas the anticipated dose for Cetrorelix is about 30 mg. Along with other attempts, the poorly soluble pamoate salt of Cetrorelix in defined particle sizes was tested for a prolonged duration of action. In rats, an effective and long lasting testosterone suppression was observed with pamoate-particles after single sc. doses of 0.5 mg/kg and 1.5 mg/kg as compared to the acetate salt. Testosterone measurement after injection of 0.5 mg/kg revealed a complete suppression for about 15 days. Using the higher dose of 1.5 mg/kg the testosterone was completely suppressed for at least 30 days. Compared to the acetate salt the duration of action was extended by 5 days (0.5 mg) and 15 days (1.5 mg), respectively.

Clinical evaluation of the Cetrorelix pamoate salt

Local tolerability of the Cetrorelix-pamoate formulation was tested in toxicological studies after intramuscular and subcutaneous injections into dogs. The results revealed no clinical signs of local intolerabilities for both the intramuscular and subcutaneous administration route and also applies for the control animals which had received the suspension medium only. In none of the animals systemic toxic effects occurred. Thus this formulation was tested in four placebo controlled Phase-I studies in healthy male individuals by intramuscular injections. When single doses of 30 and 60 mg were tested, mean plasma testosterone levels beneath 1 ng/ml were observed within 12 hours after administration. A dose related duration of testosterone suppression was found reaching 18-24 days in some but not all volunteers. On the basis of these results and the previous experience with Cetrorelix acetate, the duration of action of Cetrorelix pamoate should also be investigated for the purpose of a maintenance therapy after an initial 5-days pre-treatment with Cetrorelix acetate (so called "loading dose concept") . In the corresponding study treatment started with daily sc. injections of 10 mg Cetrorelix acetate for 5 days followed by a single administration of 60 mg Cetrorelix pamoate i.m. on day 6. Using this treatment schedule a reliable maintenance of testosterone-suppression to castration level could be achieved. To further simplify the "loading dose concept"" consequently, a further study has been performed using only 2 administrations of Cetrorelix pamoate, 24 h apart from each other. From all tested dose schedules the highest dosage group of 60 mg +60 mg showed the most pronounced duration of action. Testosterone was suppressed to castration levels for 25 up to 39 days.

The i.m. route of administration was clearly well tolerated. Decreased libido and hot flushes were reported in 2 of 5 subjects after 22 days of testosterone suppression. Recently, first clinical trials in patients suffering from uterine myoma have been initiated and first results indicate an effective reduction of myoma size during treatment with Cetrorelix. The Cetrorelix plasma level measurements in these different studies revealed a relatively high variability especially within the first 8 days after administration and hence this formulation has to be further improved.

On the basis of the preclinical results and the clinical data from the treatment of more than 600 individuals, Cetrorelix proved to be a safe LHRH antagonist and because of its hormone-suppressive effectiveness has the potential to be applied in a variety of clinical conditions. Since the antagonist effectively inhibits spermatogenesis and ovulation it might, in addition be useful in male contraception and in the prevention of gonadal damage by cytostatic agents during chemotherapy. Presently daily injections of the lyophilisate are used. However, for long term treatment a depot formulation is desirable and different approaches towards a clinically suitable depot formulation are under evaluation.

The results described indicate that the pamoate salt of Cetrorelix has the effect of a slow release formulation and a suitable release characteristic, although the observed high inter-subject variability needs to be improved. Beside the indications of hormone dependent cancers, benign indications as BPH, uterine fiboids and endometriosis will be investigated.

References

1. Matsuo, H., Baba, Y., Nair, R.M.G., Arimura, A. and Schally, A.V., Biochem. Biophys. Res. Commun., 43 (1971) 1334.
2. Schmidt, F., Sundaram, K., Thau, R.B. and Bardin, C.W., Contraception, 29 (1984) 283 .
3. Bajusz, S., Kovacs, M., Gazdag, M., Bokser, L., Karashima, T., Csernus, V.J., Janaky, T.Guoth, J. and Schally, A.V., Proc. Natl. Acad. Sci.,U.S.A., 85 (1988) 1637.
4. Behre, H.M., Bckers, A., Schlingheider, A. and Nieschlag,E., Clin. Endocrinol., 40 (1994) 241.
5. Behre, H.M., Kliesch, S., Phse, G. and Nieschlag, E., Exp. and Clin. Endocrinol., 102 (1994)53.
6. Gonzalez-Barcena, D., Vadillo-Buenfil, M., Gomez-Orta, F., Fuentes-Garcia, M., Cardenas-Cornejo, I., Graef-Sanchez, A., Comaru-Schally, A.M. and Schally, A.V., The Prostate, 24 (1994) 84.
7. Reissmann, T., Felberbaum, R., Diedrich, K., Engel, J., Comaru-Schally, A.M. and Schally, A.V., Human Reprod. 10 (1995) 1974.
8. Diedrich, K., Diedrich, C., Santos, E., Zoll, C., Al-Hasani, S., Reissmann, T., Krebs, D. and lingmler,D., Human Reprod., 9 (1994) 788.
9. Leroy, I., D'Acremont, M.F., Brailly-Tabard, S., Frydman, R., DeMouzon, J. and Bouchard, P., Fertil. Steril., 62 (1994) 461.

Characterization and identification of growth hormone receptors in snakeheaded fish (*Ophiocephalida argus* cantor) liver membrane

Xun Sun[a], Shang-Quan Zhu[b] and Christopher H. K. Cheng[b]
[a]*Shanghai Institute of Biochemistry, Chinese Academy of Sciences, Shanghai 200031, China*
[b]*Department of Biochemistry, The Chinese University of Hong Kong, Hong Kong*

Introduction

Growth hormone (GH) initiates diverse responses in various target tissues by binding to its specific receptor located on the plasma membrane of cells [1]. In contrast to the extensive characterization of growth hormone receptors (GHR) in mammalian tissue, little information is available concerning the properties and distribution of GHR in fish [2], especially the freshwater fish in China. We have screened the concentration of liver GHR of freshwater fish in China and found that the snakeheaded fish liver contained higher concentration of GHR than those of other fish. The present study was conducted to characterize and identify GHR in snakeheaded fish liver membrane (SFLM).

Materials

Recombinant bream growth hormone (brGH) and recombinant human growth hormone (hGH) were obtained from Bresatec Limited, Australia. Snakeheaded fish was purchased from a local market, and the liver membrane was prepared according to the method of Cuatrecasas [3].

Results and Discussion

The association of ^{125}I-brGH to SFLM was a time and temperature dependent process. At 25°C, a steady state was obtained in 16 h and was maintained for about 40 h (Fig. 1). Ta(1/2) was found to be 106 minutes. Dissociation of the hormone-receptor complex was very slow, about 30% of bound ^{125}I-brGH could be dissociated after incubation at 25°C for 22 h in the presence or absence of brGH.

The binding of ^{125}I-bGH to SFLM was saturated with increasing amounts of membrane in the presence of fixed amount of labeled brGH (Fig. 2).

The specific binding of ^{125}I-brGH to SFLM was inhibited by unlabeled brGH in a dose-dependent manner with IC50 13.4 ng/ml but was not affected by insulin. Human GH also partially competed for binding of ^{125}I-brGH but with a 150-fold reduction in potency as compared with brGH (Fig. 3).

When the ratio of bound to free ^{125}I-brGH was plotted as a function of bound brGH, a straight plot was obtained (Fig. 3). The equilibrium constant of association was 1.45± 0.23 10^9M^{-1} and the binding capacity was 198± 57 fmol/mg protein.

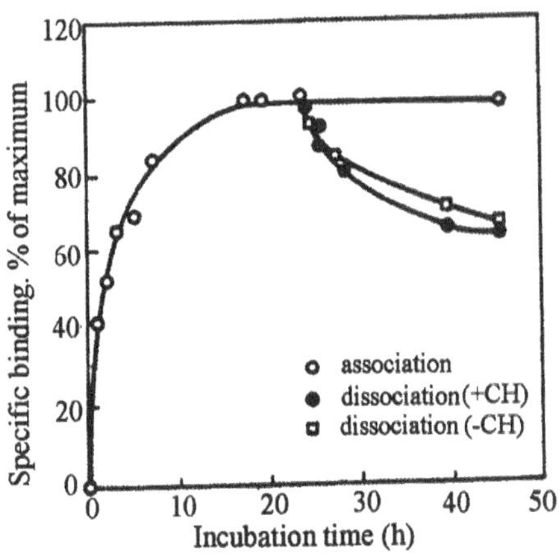

Fig. 1. Time course of association and disso-ciation of ^{125}I-brGH to/from SFLM. Each point is the mean of triplicate determination.

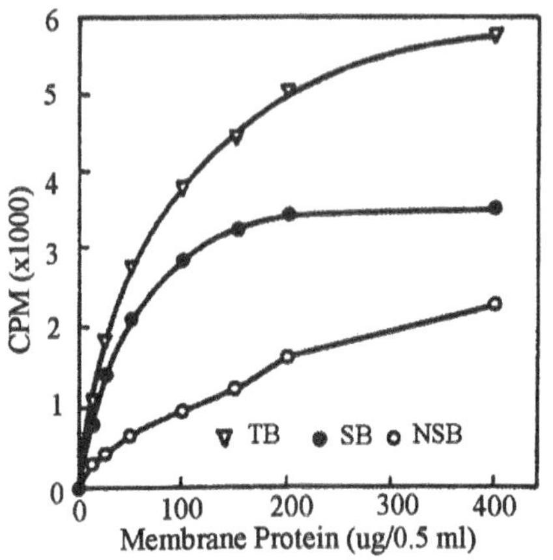

Fig. 2. Effect of membrane protein concentration on total (TB), nonspecific (NB) and specific (SB) binding of ^{125}I-brGH to SFLM in presence of fixed amount of labeled br-GH. Each point is the mean of triplicate determination.

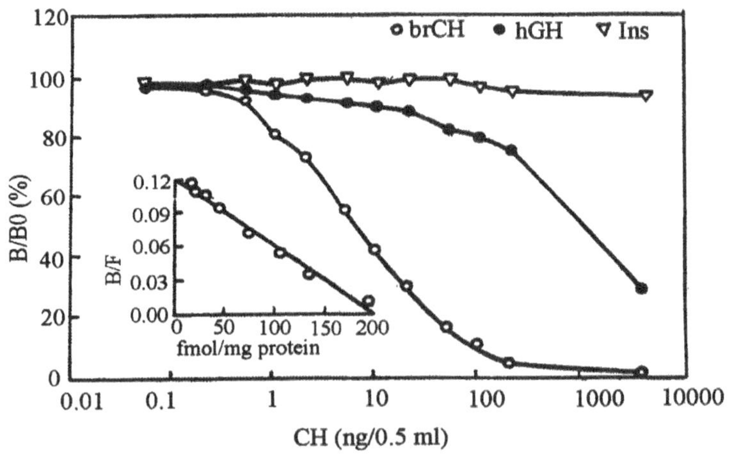

Fig. 3. Competition for specific binding of ^{125}I-brGH to SFLM by unlabeled br-GH, human GH and insulin. Each point is the mean of three determinations. Inset: Scatchard plot.

Different molecular weight forms of GHR on SFLM were observed by the following three methods.

1. Sephadex G-200 column chromatography. Triton X-100 solubilized SFLM was incubated to equilibrium with ^{125}I-brGH, and the mixture was applied onto a Sephadex G-200 column (Fig. 4). The peak of ^{125}I-brGH-receptor complex corresponded to 200-400 kD.
2. Western blot. Mouse anti-rabbit GHR monoclonal antibody was used to carry out Western blot. A single band with a molecular weight of about 90-93 kD was detected.
3. Crosslinking. After SFLM was incubated to equilibrium with ^{125}I-brGH, disuccinimidyl suberate was added to crosslink ^{125}I-brGH-receptor complex. Two major autoradiographic bands corresponding to Mr 65 and 89 kD were observed on SDS-PAGE.

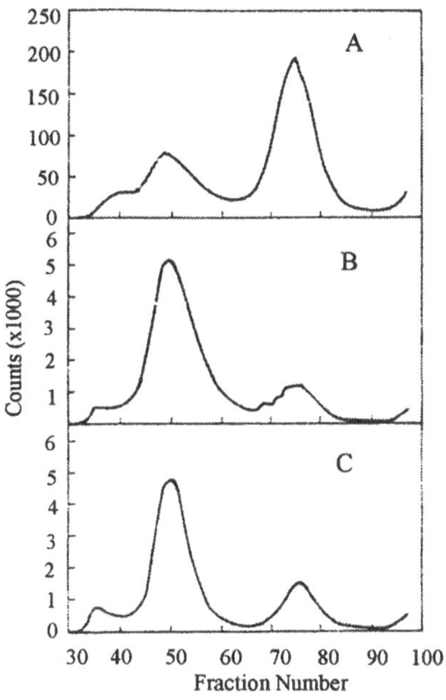

Fig 4. Gel filtration patterns demo-nstrating specific and irreversible binding between ^{125}I-brGH and solu-bilized SFLM.
A. 0.5ml solubilized membrane rece-ptor was incubated with ^{125}I-brGH overnight at RT., cooled in ice and applied onto a column (1x100cm) of Sephadex G-200. The column was equilibrated and run at 4°C with 50 mM Tris-HCl, pH 7.2 containing 0.1 M NaCl, 10 mM $CaCl_2$ and 0.1% Triton X-100. The flow rate was about 6ml/hr, and each fraction contained about 0.9ml;
B. 0.5ml of the first peak in A was rechromatographed;
C. Another 0.5ml of the first peak in A was incubated with 100 g unlabeled br-GH overnight at RT., then reapplied onto G-200.

The above results may be explained by the partial hydrolysis and aggregation of GHR in the experimental process. Similar phenomenon was observed in earlier study on mammalian GHR [4].

References

1. Postel-Vinay, M.C. and Finidory, J., Eur. J. Endocrinol., 133 (1995) 654.
2. Yao, K., Nice, P.D., Gac, F.L. and Bail, P.Y.L., Generaland Comp. Endocrinology., 81 (1991) 72.
3. Cuatrecasas, P., Proc. Nat. Acad. Sci. U.S.A., 69 (1972) 318.
4. Hoequette, J.A., Postel-Vinay, M.C., Djiane, J., Tar, A. and Kelly, P.A., Endocrinology, 127 (1990) 1665.

Effect of divalent cations and disulfide reducing agents on the specific binding of growth hormone to snakeheaded fish liver membrane

Xun Sun[a], Shang-Quan Zhu[a] and Christopher H. K. Cheng[b]
[a]Shanghai Institute of Biochemistry, Chinese Academy of Sciences, Shanghai 200031, China
[b]Department of Biochemistry, The Chinese University of Hong Kong, Hong Kong

Introduction

The interaction of hormones with their receptors is affected by many factors, such as temperature, pH, ions etc [1]. Here, we report the effect of divalent cations and disulfide reducing agents on the binding of bream growth hormone (brGH) to snakeheaded fish liver membrane (SFLM).

Results and Discussion

The divalent ions could markedly increase the binding of ^{125}I-brGH to SFLM and the optimum concentration of ions was found to be in the range of 8-12 mM in which Ca^{2+}, Mg^{2+} and Mn^{2+} could increase the specific binding to 230%, 180% and 200% of that in the absence of ion, respectively (Fig. 1).

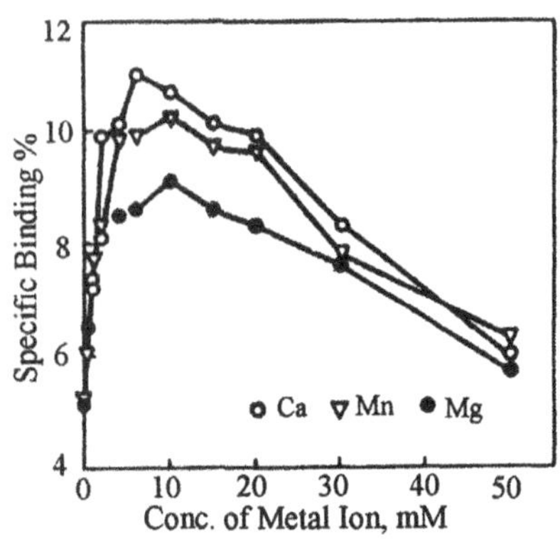

Fig. 1. Dose-response effect of divalent cation on ^{125}I-brGH binding to SFLM. Each point represents the mean of triplicate determination.

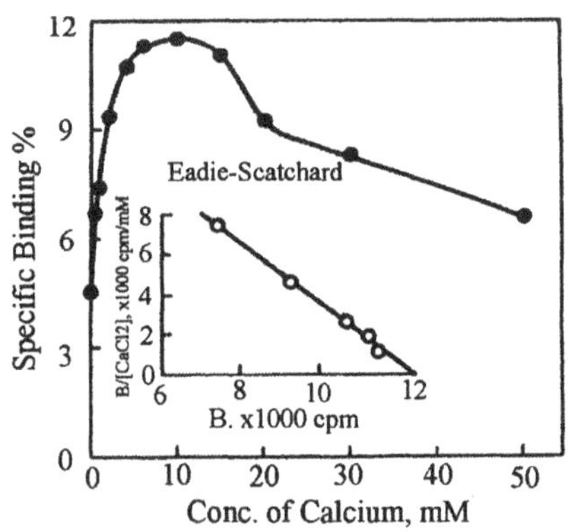

Fig. 2. Eadie-Scatchard analysis for influence of Ca^{2+} concentration on ^{125}I-brGH binding to SFLM. Each point represents the mean of triplicate determination.

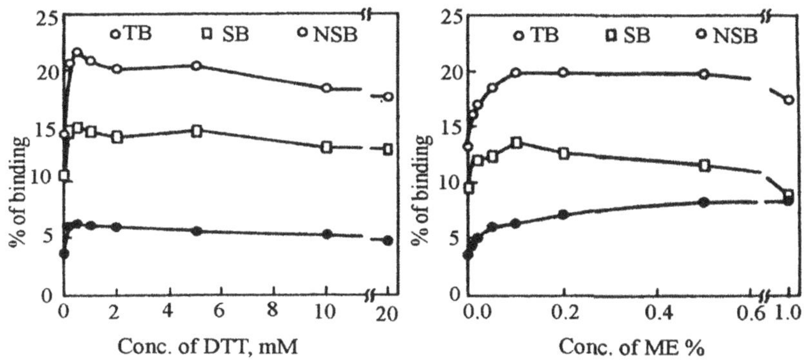

Fig. 3. Effect of reducing agents on binding of ^{125}I-brGH to SFLM.
A: in the presence of 0-20mM dithiothreitol (DTT);
B: in the presence of 0-1% beta-mercaptoethanol (ME).
Each point represents the mean of triplicate determination.

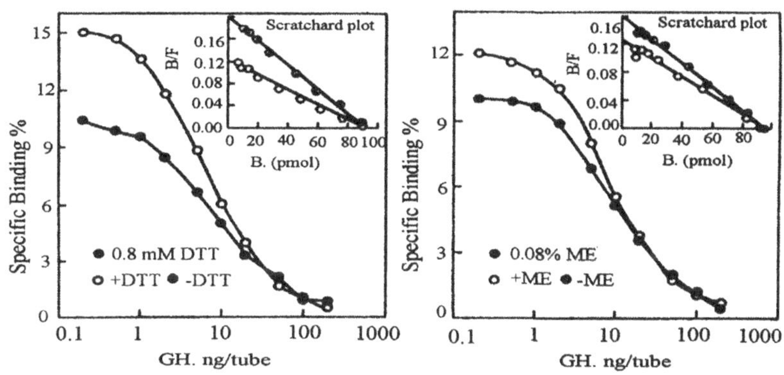

Fig. 4. Scatchard plot analysis of specific binding of ^{125}I-brGH to SFLM
A: in the presence or absence of 0.8mM DTT;
B: in the presence or absence of 0.08% ME.
Each point represents the mean of triplicate determination.

Using Eadie-Scatchard plot for dynamic analysis of calcium binding, it was found that in the brGH-receptor complex there was a low affinity binding site for calcium with a Km of 0.384 ± 0.018 mM (Fig 2),when calcium concentration was in the range of 0.1-10 mM. In addition, Scatchard plot showed that the affinity constant for ^{125}I-brGH binding to SFLM changed from $1.04 \pm 0.16 \times 10^9$ M^{-1} to $1.29 \pm 0.31 \times 10^9$ M^{-1} (n=3, P<0.01) in the presence of 10 mM $CaCl_2$.

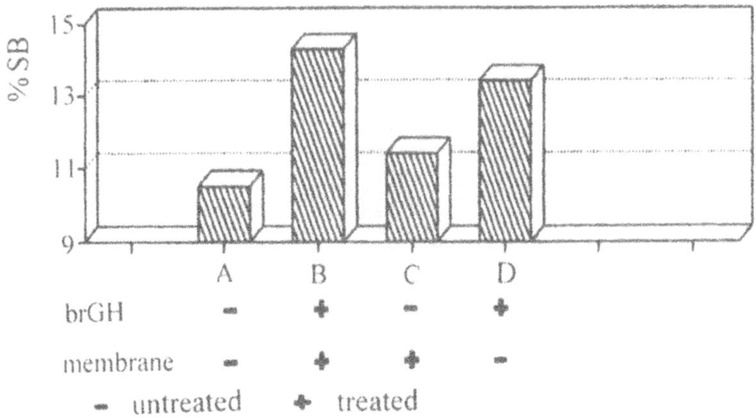

Fig. 5. Effect of pretreatment of brGH, membrane or both with 0.8 mM DTT on specific binding. A vs B, A vs D, P<0.01, A vs C, P<0.05. Each value represents the mean of triplicate or more determinations.

Addition of 0.1-20 mM dithiothreitol (DTT) or 0.01-1% mercapto-ethanol (ME) to receptor binding system caused a significant dose-dependent increase in the binding of ^{125}I-brGH to SFLM (Fig. 3).The Scatchard analysis showed that in the presence of 0.8 mM DTI or 0.08% ME, the binding affinity constant was changed from $1.26\pm0.12 \times 10^9$ M^{-1}(in the absence of reductants) to $1.62\pm0.15 \times 10^9$ M^{-1} or $2.18\pm0.23\times10^9$ M^{-1} respectively. However, the binding capacity remained unchanged (Fig. 4). From Fig. 5, it was shown that DTT increased the interaction between brGH and its receptor on SFLM, suggesting that this reaction may include disulfide bond exchanges [2-4].

References

1. Rowlinson, S.W., Barnard, R., Bastinas S., Robins, A. J., Wells, J. Q. E., Brinkworth, R., and Waters., M J., Biochemistry, 33 (1994) 11724.
2. Gray, E. S., and Tsai, R. W., J. Exp. Zool., 268 (1994) 428.
3. Jarret, L., and Smith, R. M., Proc. Natl. Acad. Sci. USA, 80 (1983) 1023.
4. Moore, W. V., Wohnilich, L. P. and Fix, J. A., Endocrinology., 112 (1983) 2152.

Molecular mechanisms of apoptosis induced by CD3ε peptide

De-Xian Zheng[a], Shi-Lian Liu[a], Bao-Ping Wang[b] and Cox Terhorst[b]
[a]*Institutes of Basic Medical Sciences and Hematology, CAMS, Beijing 100005, China*
[b]*Harvard Medical School, Boston, MA 02215, U. S. A.*

Introduction

The present study was conducted to investigate the mechanisms of thymus shrinkage in transgenic mice with CD3ε peptide overexpression.

Results and Discussion

1. TUNEL assay [1] showed that CD3ε overexpression facilitated cell death in thymocytes of transgenic mice (Fig. 1).
2. In order to understand the mechanisms of apoptosis in the transgenic mice with CD3ε overexpression, a chimera that fused the extracellular and transmembrane domains of human CD8 to the cytoplasmic domain of CD3ε was synthesized (Fig. 2) and stably expressed in CD8$^-$ Jurkat T lymphocytes (CD8$^+$) after transfection. The untransfected (Fig. 3A) and pcDNA3 plasmid DNA transfected (Fig. 3B) Jurkat cells grew with no cell death, but the CD8ε/pcDNA3 transfected cells grew with 33% of cell death (Fig. 3D) after stimu-

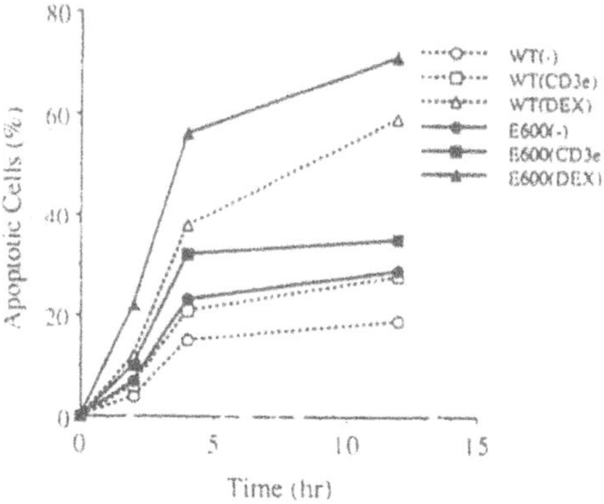

Fig. 1. Analysis of apoptosis in the thymocytes of transgenic mice

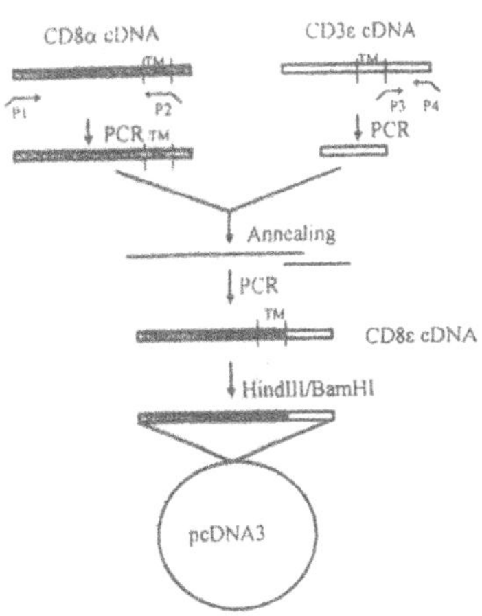

Fig. 2. Diagram of the construction of chimeric cDNA (CD8ε) and the eukaryotic expression vector.

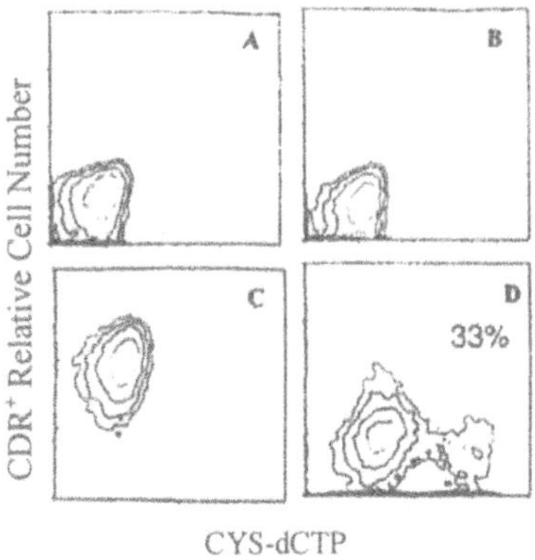

Fig. 3. Apoptosis analysis
A: Jurkat cells ;
B: pcDNA3 plasmid DNA transfected cells;
C: Negative control;
D: CD8ε transfected Jurkat cells;

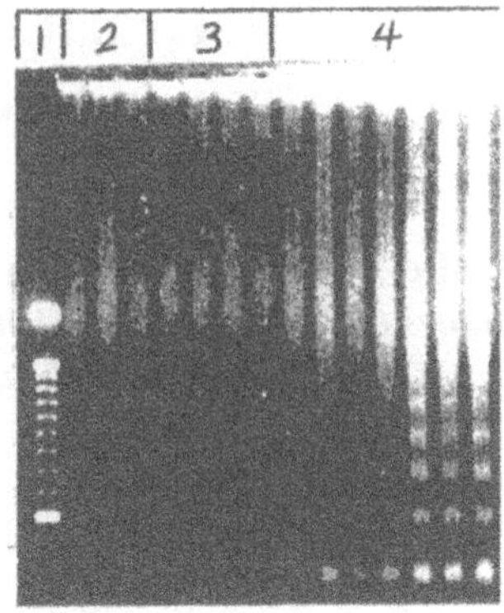

Fig. 4. DNA fragmentation.
1. DNA molecular marker;
2. Jurkat cell;
3. pcDNA3 transfected cells;
4. CD8ε transfected cells.

lating the cells with anti-CD8 antibody, indicating that the ITAM [2] in the CD3ε cytoplasmic tail can couple the external stimulating signals.

3. A typical DNA ladder occurred in the McAb-treated JK/CD8$^+$ cells but not in the control cells (Fig. 4), indicating that JK/CD8$^+$ cells died by apoptosis. From this interesting finding we propose a hypothesis that there might exist a death motif in the cytoplasmic tail of CD3ε.

References

1. Gavrieli, Y., Sherman Y. and Ben-Sasson S. A., J. Cell Biol., 119 (1992) 493.
2. Cambier, J.C., Immunol. Today,. 16 (1995) 110.

A common antigenic peptide of several HCV proteins

Jia-Xi Xu
Department of Chemistry, Peking University, Beijing 100871, China

Introduction

Hepatitis C virus (HCV), a member of the *Flaviviridae* family of positive-strand RNA viruses, is the major etiologic agent of vital non-A, non-B hepatitis. Numerous studies have indicated that most HCV-infected individuals will become chronic carriers, a stage that may lead to development of cirrhosis and hepatocellular carcinoma [1,2]. Thus, it is very important to develop a specific and sensitive diagnosis for the detection of HCV antibodies at an early phase of HCV infection and an efficient anti-HCV vaccine in HCV-infected control. Geographically distinct isolates have been molecularly cloned and sequenced. Sequence analysis revealed that the HCV genome is approximately 9400 nucleotides long, encoding a 3011-amino acid polyprotein [3]. Up to now, many sequences of putative HCV proteins have been reported. Among these proteins, envelope protein and envelope protein E1, core envelope protein, genome polyprotein and precursor protein may serve as important targets for anti-HCV vaccine development, since they are likely to be involved in host range immune selection, virus neutralization and persistent infection properties of virus [4]. In order to determine the most immunogenic regions, after analyzing the sequences of several putative HCV proteins, a common antigenic peptide shared by these putative HCV proteins was synthesized. This peptide may be valuable for studies of HCV diagnosis and vaccine development.

Results and Discussion

Prediction of epitopes

The protein sequences of putative HCV proteins mentioned above, envelope protein and envelope protein E1, core envelope protein, genome polyprotein and precursor protein were obtained from published studies [5-10]. Their hydrophilicities, accessibilities, and secondary structures were predicted by using a PC-GENE program with the aid of computer. After analyzing the sequences of putative HCV proteins, a decapeptide shared by these putative HCV proteins, H-Gly-Asn-Ser-Ser-Arg-Cys-Trp-Val-Ala-Leu-OH, was selected as a common antigenic peptide. It is located in 42-51 amino acid residues of envelope protein, 37-46 of envelope protein E1, 74-83 of precursor protein, 233-242 of both core envelope protein and genome polyprotein.

Synthesis of common epitopic peptide

The common epitopic peptide was synthesized by a solid phase FMOC-chemistry strategy on an ABI model 431A automatic peptide synthesizer (Applied Biosystems, Inc., Foster City, CA, USA) according to the manufacturer's protocols. After purification on a Sephadex G15 column and high performance liquid chromatography, characterization by amino acid analysis and mass spectrum, peptide was used directly in enzyme-linked immunosorbent assay (ELISA). This peptide may be valuable for studies of HCV diagnosis and vaccine development.

Acknowledgment

Project (29772002) was supported by National Natural Science Foundation of China.

References

1. Hollinger, F. B., Non-A, Non-B hepatitis viruses, In Fields, B. N. and Knipe, D. M.(eds.) Virology Raven Press, New York, 1990, p.2239.
2. Kiyosawa, K., Sodeyama, T. and Tanaka, E., Hepatology, 12 (1990) 671.
3. Simmonds, P., Holmes, E. C. and Cha, T. H., J. Gen. Virol., 74 (1993) 661.
4. Mink, M. A., Benichou, S. and Madaule, P., Virol., 200 (1994) 246.
5. Houghton, M. A., Weiner, A. J. and Han, J.,Hepatology, 2 (1991) 381.
6. Hijikata, M., Kato, N. and Ootsyama, Y., Proc. Natl. Acad. Sci. USA, 88 (1991) 2451.
7. Takamizawa, A., J. Virol., 65 (1991) 1105.
8. Weiner, A. J., Brauer, M. J. and Rosenblatt, J.,Virol., 180 (1991) 842.
9. Hijikata, M., Kato, N. and Ootsyama, Y., Biochem. Biophys. Res. Commun., 175 (1991) 220.
10. Weiner, A. J., Geysen, H. M. and Christopherson, C., Proc. Natl. Acad. Sci. USA, 89 (1992) 3468.

Multivalent synthetic peptide vaccine against schistosomiasis: Preparation of multiple antigenic peptide

Jia-Xi Xu
Department of Chemistry, Peking University, Beijing 100871, China

Introduction

Schistosomiasis is a chronic debilitating disease affecting humans in 70 countries all over the world, ranking second in parasitoses next to malaria. It is estimated that 200-300 million people are infected by various species of schistosomes. In spite of the development of active and safe drugs, prevention of rapid reinfection has remained a serious problem, requiring repeated drug applications over a long period. There is a clear and obvious reason therefore for an effective vaccine including significant levels of protection against the invasive stage of the parasite [1].

Among the candidates for an anti-schistosomal vaccine, the glutathione S-transferases (GSTs) are of major immunological interest. These GST isoenzymes appear to play a key role in the parasite detoxification system. Thus, the glutathione S-transferases have potential as components of a vaccine against schistosomiasis [2,3]. Immunochemical analyses have revealed that both schistosoma mansoni and schistosoma japonicium have at least two GST isoenzymes, of 26 and 28kDa (Sm26, Sm28, Sj26 and Sj28). Sm26 has two different types (termed Sm26/1 and Sm26/2), which share an 82% identity at amino acid level [4]. In previous reports [5-7], the antigenic peptides of three 26kDa GST isoenzymes (Sj26, Sm26/1 and Sm26/2) were studied. Three high antigenic peptides were derived from 26 kDa GSTs of *schistosoma mansoni* and *japonicium*, potential as components of a vaccine against Schistosomiasis. In this paper, three multiple antigenic peptides (MAPs) were synthesized by using the above three effective peptides as antigenic peptides especially, Lys-tree as core with eight antigenic peptides.

Results and Discussion

Prediction of epitopes

Nucleotide sequences and amino acid sequences of the 26kDa GST isoenzymes (Sj26, Sm26/1 and Sm26/2) were obtained from published data [4]. Their epitopes were predicted according to their hydrophilicity (Hopp and Woods method, HPLC retention method), flexibility and accessibility, secondary structure and charge distribution by a computer program Goldkey [7-9]. The secondary structure of protein was predicted by the Chou and Fasman method [10].

Synthesis of epitopic peptides

Epitopic peptides and MAPs were synthesized by a manual solid phase Boc-chemistry strategy [11]. After HF-cleavage and purification on a Sephadex G25 column, the purification and characterization were carried out with HPLC and amino acid analysis.

Peptides were used directly in dot enzyme-linked immunosorbent assay (ELISA).

The results of dot blot immunobinding assay given in Table 1, showed that effective antigenic peptides are located in the middle and near C-terminus of 26kDa GSTs. Moreover, the epitopic peptides with high antigenicity are located near C-terminus of 26kDa GSTs. Three peptides, J5, S5 and M4 among them showed high antigenicity. They are potential peptide antigens of a vaccine against schistosomiasis.

Table 1. Antigenicities of epitopic peptides of 26kDa GSTs

Epitopic peptides	A6McAb	Anti-Sj-IgG
J5(SjP187-202)	++	++
J6(SjP206-218)	+	+
S3(Sm/1P105-120)	-	+++
S5(Sm/1P187-202)	+++	+++
M4(Sm/2P187-202)	++	++

Preparation of multiple antigenic peptide (MAP)

According to Table 1, three highly antigenic peptides derived from 26 kDa GSTs of *schistosoma mansoni* and *japonicium*, potential as components of a vaccine against Schistosomiasis were obtained. Three multiple antigenic peptides (MAPs) were synthesized by using the above three effective peptides as antigenic peptides especially, Lys-tree as core with eight antigenic peptides [12]. The structures of MAPs and the sequences of these three peptides are as follows.

MAP: β-Ala-Lys-$(Lys)_2$-$(Lys)_4$-$(antigenic\ peptide)_8$

Antigenic peptide: PQIDKYLKSSKYIAWP
PPIKNYLNSNRYIKWP
PNIKNYLNSSKYIKWP

Among various strategies for immunization, the strategy of using synthetic peptides is perhaps the most appealing, both in conceptual simplicity and in practical convenience, as recounted by Tam [13]. It is an important approach that uses a peptide segment with one or a few of the desired epitopes of a protein or whole organism, both are many times larger than the peptide. By keeping only the essential epitope(s) and eliminating all other unwanted portions of the protein or organism, a synthetic peptide immunogen provides the desired selectivity theoretically. Up to now, the use of synthetic peptides has become a well accepted method for the preparation of site-specific antibodies for various biochemical and functional studies [14,15].

As we have known, peptides are generally poor immunogens and are eliminated from the body rapidly. For immunization purposes, peptides have to be modified and administered with a formulation containing an adjuvant and a vehicle, such as paraffin oil. Many different approaches for peptide modification and presentation have been described in recent years. These include the traditional approaches of conjugation to a protein carrier and polymerization, as well as the new approach known as the multiple antigen peptide

(MAP) which contains well defined structures with or without a built-in adjuvant. The MAP approach developed by Dr. James P. Tam replaces the protein carrier with a small score matrix comprising oligomeric lysine; it is a unique presentation system that provides peptide epitopes unambiguously in multiple copies. As a result, a MAP provides a very high density of the desired peptide epitopes at the surface of the construction with a small non-protein core matrix as a scaffold. This eliminates many of the disadvantages associated with the use of a protein carrier. Another advantage is that the following two steps are unnecessary in preparation of peptide immunogen, to select a carrier and to select a conjugation method to prepare the peptide-carrier complex [16].

In summary, the antigenic peptides of three 26kDa GST isoenzymes (Sj26, Sm26/1 and Sm26/2, potential as components of a vaccine against Schistosomiasis) were studied. Three high antigenic peptides were derived from 26 kDa GSTs of *schistosoma mansoni* and *japonicium*. Three multiple antigenic peptides (MAPs) were synthesized by using these three effective peptides as antigenic peptides especially, Lys-tree as core with eight antigenic peptides. The animal experiments of immunogenicity of these three MAPs are in progress.

Acknowledgment

Project (29772002) was supported by National Natural Science Foundation of China.

References

1. Balloul, J. M., Sondermeyer, P. and Dreyer, D., Nature, 326 (1987) 149.
2. Henkel, K. L., Davern, K. M. and Wright, M. D.,Mol. Biochem. Parasitol., 40 (1990) 35.
3. Trottein, F., Kieny, M. P., Verwaerde, C., ibid., 41 (1990) 41.
4. Wright, M. D., Harrison, R. A.and Melder, A. M., ibid., 49 (1991) 177.
5. Xu, J. X. and Cai, M. S., In Lu, G. S., Tam, J. P. and Du, Y. C.(eds.) Peptides: Biology and Chemistry (Proceedings of the 1994 Chinese Peptide Symposium), ESCOM, Leiden, 1995, p. 245.
6. Xu, J. X. and Cai, M. S., Ibid., p. 256.
7. Xu, J. X., Cai, M. S. and Shi, Y. E., Chem. J. Chinese Univ., 17 (1996) 424.
8. Hopp, T. P. and Woods, K. R., Proc. Natl. Acad. Sci. USA, 78 (1981) 3824.
9. Janin, J., Nature, 277 (1979) 491.
10. Chou, P. Y. and Fasman, G. D., Biochem., 13 (1974) 222.
11. Xu, J. X. and Jin, S., Chin. Chem. Letts., 4 (1993) 213.
12. Tam, J. P., Proc. Natl. Acad. Sci. USA, 85 (1988) 5409.
13. Tam, J. P. in Wisdom, G. B. (ed.) Peptide antigens--A practical approach, Oxford University Press, Oxford-New-York-Tokyo, 1994, p. 83.
14. Ada, G. L., The Lancet, 355 (1990) 523.
15. Brown, F., The Lancet, 355 (1990) 587.
16. Wolowuczuk, I., Auriault, C. and Bossus, M., J. Immunol., 146 (1991) 187.

Synthesis of antigenic peptides for HEV diagnosis

Jia-Xi Xu[a], Zhi-Hui Qin[b] and Meng-Shen Cai[b]
[a]*Department of Chemistry, Peking University, Beijing 100871, China*
[b]*School of Pharmaceutical Sciences, Beijing Medical University, Beijing 100083, China*

Introduction

Hepatitis E virus (HEV) is a major causative agent of enterically transmitted non-A, non-B hepatitis. Outbreaks of HEV infection have occurred in many developing countries since the first known HEV outbreak in 1955 in New Delhi, India. HEV is an especially serious problem in these countries because of a high fatality rate among pregnant women [1]. The diagnosis of Hepatitis E was previously made by serologic exclusion of other cases of acute viral hepatitis. The cloning and sequencing of the HEV genome have led to the development of prototype immunoassays for the serologic detection of HEV infection [2]. Using the sequence of the HEV genome of Chinese strain containing three open reading frames (ORFs), three antigenic peptides have been derived from HEV protein of Chinese strain, two in structural protein components which encode by ORF2 and one in unknown function protein which encodes by ORF3 [3,4]. These three peptides were synthesized and used successfully for the analysis of serum samples of HEV infected patients.

Results and Discussion

Prediction of epitopes

Nucleotide sequences and amino acid sequences of the HEV genome of Chinese strain (CE 1.1) were obtained from published data [3,4]. The epitopes, NS33, S30, S42, were predicted according to their hydrophilicity, flexibility and accessibility by a computer program Goldkey [5,6]. The secondary structure of protein was predicted by Chou and Fasman method [7]. The selected synthetic peptides contain hydrophilic, flexible and accessible sequences represented in the predicted secondary structure as β-turns or random coils. NS33 of the synthetic peptides is predicted to have a potential to be transmembrane α-helices in ORF3. The function of the ORF3 protein is unknown.

A few experimental data have indicated that the C-terminal region of the ORF2 protein contains antigenic epitopes [1-4,8,9]. Two peptides, S30 and S42, were selected from the hydrophilic and flexible region of the N-terminal half of the protein for synthesis with the knowledge that internal proteins of viral particles such as hepatitis B, C, D viruses have been important diagnostic reagents.

Synthesis of epitopic peptides

Epitopic peptides were synthesized by a solid phase Fmoc-chemistry strategy on an ABI model 431A automatic peptide synthesizer (Applied Biosystems, Inc., Foster City, CA, USA) according to the manufacturer's protocols. After purification on a Sephadex G25 column and characterization by amino acid analysis, high performance liquid chromatography, peptides were used directly in enzyme-linked immunosorbent assay (ELISA).

Antigenicity of synthetic peptides

Reactivity of each of the synthetic peptides was determined by ELISA using serum specimens from patients with hepatitis E. All the specimens were initially tested by Western blot. All synthetic peptides were reactive with the serum specimens that were anti-HEV positive.

Diagnostic relevance of synthetic peptides

A set of peptides composed of these three synthetic peptides was used to analyze serum samples obtained from a recent outbreak in Xinjiang, China. These serum samples included 111 serum samples of acute HEV patients, 32 serum samples of convalescent HEV patients, 113 serum samples of sporadic HEV patients and 106 serum samples of HEV infectious macaques. All HEV serum samples were identified by Western blot assay. The results are shown in Table 1. Our results were compared with those of diagnostic kits of American Genelabs. Our positive samples which were negative in Genelabs kit were confirmed by neutralization tests and protein print tests.

The coincidence rates, sensitivities and specificities among our kit and American Genelabs kit were compared by detections to several groups of serum samples. The results are shown in Tables 2 - 4.

According to the results of detection to HEV infectious macaque serum samples, both kits showed very good specificities and false positive rates were below 11.4%.

Table 1. Detection results of serum samples of HEV patients and HEV infectious macaques

	Sensitivity	
Serum sample of HEV	Our kit (%)	Genelabs kit (%)
Acute patient	86.5	73.9
Convalescent	6.3	0.0
Sporadic	6.2	4.4
HEV infectious	55.7	42.5
Total	45.2	36.2

Table 2. Comparative testing of our kit and Genelabs kit for macaque serum samples

	Genelabs kit		
	+	-	Total
Our kit +	11	11	22
Our kit -	4	31	35
Total	15	42	57

Coincidence Rate: (11+31)/57=73.68%
Genelabs kit to our kit
Sensitivity: 11/22= 50% Specificity: 31/35=88.57%
False Positive Rate: 4/35=11.43% False Negative Rate: 11/22=50%

Table 3. Comparative testing of our kit and Genelabs kit for patient serum samples

	Genelabs kit		
	+	-	Total
Our kit +	18	4	22
Our kit -	0	7	7
Total	18	11	29

Coincidence Rate: (18+7)/29=86.2%
Genelabs kit to our kit
Sensitivity: 18/22= 81.8% Specificity: 7/7=100.0%
False Positive Rate: 0/7=0.00% False Negative Rate: 4/11=36.4%

Table 4. Comparative testing of our kit and Genelabs kit for patients and macaque serum samples

	Genelabs kit		
	+	-	Total
Our kit +	29.	15	44
Our kit -	4	38	42
Total	33	53	86

Coincidence Rate: (29+38)/86=77.91%
Genelabs kit to our kit
Sensitivity: 29/44= 65.91% Specificity: 38/42=90.48%
False Positive Rate: 4/42=9.52% False Negative Rate: 15/44=34.09%

From Tables 3 and 4, our kit and Genelabs kit showed good coincidence rate and specificity. However, Genelabs kit showed a lower sensitivity and a higher false positive rate than our kit.

In summary, an enzyme-linked immunosorbent assay (ELISA) for hepatitis E virus (HEV) was prepared by using synthetic antigenic peptides NS33, S30 and S42 derived from the protein of HEV. It has high specificity and sensitivity and is capable of detecting specific antibodies in the acute, convalescent, sporadic phase of HEV infection. Our ELISA diagnostic kit for anti-HEV is better than Genelabs kit. It has a higher specificity and sensitivity, lower false positive rate than Genelabs kit. The reason is that American Genelabs kit is composed of two antigenic peptides prepared by cloning and expression of HEV gene recombinant with poor purity. Therefore it is less sensitive. However, our ELISA kit is composed of three artificial synthetic HEV antigenic peptides which include all three epitopes of HEV. Thus it shows high sensitivity and specificity.

Acknowledgments

We thank Prof. H. Zhang and his colleagues for help in the ELISA experiment.

References

1. Khudyakov, Y.E., Khudyakova and N.S., Fields, H A., Virology, 194 (1993) 89.
2. Khudyakov, Y.E., Favorov, M.O. and Khudyakova, N.S., J. Virology, 68 (1994) 7067.
3. Zhuang, H., Li ,F. and Guo, X.X., Chinese J. Experimental and Clinical Virology, 7 (1993) 14.
4. Bi, S.L., Liu, C.B. and Cao, X.Y., Chinese J. Virology, 8 (1992) 271.
5. Hopp, T.P., Woods, K.R., Proc. Natl. Acad. Sci. USA, 78 (1981) 3824.
6. Janin, J., Nature, 277 (1979) 491.
7. Chou, P.Y. and Fasman, G.D., Biochem., 13 (1974) 222.
8. Ray, R., Jameel, S. and Manivel, V., Virology, 189 (1992) 359.
9. Tam, A.W., Smith, M.M. and Guerra, M.E., Virology, 185 (1991) 120

A very strongly acidic pentapeptide from *Panax ginseng*

Qi Li, Chong-Xu Fan, Xiao-Jing Wang, Yun-Hua Ye and Qi-Yi Xing
Department of Chemistry, Peking University, Beijing 100871, China

Introduction

Panax ginseng as famous Chinese traditional medicine attracts chemists interested in its effective chemical compositions, especially the water-soluble part. We show that oligopeptides are very important water-soluble constituents in *Panax ginseng* which play an important role in neurophysiological activity, inhibition of the lipolysis of fat etc. Until now six oligopeptides have been isolated and identified from *Panax ginseng* by our group [1-7], they are:

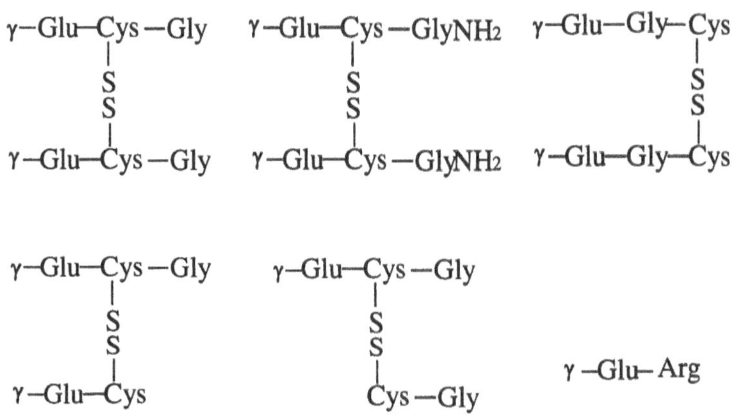

Recently, a very strongly acidic peptide was isolated; its amino acid constituent, N-terminal and C-terminal were determined.

Results and Discussion

The procedure was used to purify this pentapeptide as shown in the scheme. Most of the saponins and saccharides were removed by ion exchange resin, and all the charged peptides were collected. The crude fractions were gel filtrated through Sephadex G-15, and all the fractions which gave the positive reaction to ninhydrin were subjected to amino acid analysis. The remarkable increase of Asp, Glu and Gly indicated that a very strong acidic peptide exists in this fraction. In order to remove free Glu, Asp etc., the fraction was separated by DEAE-Sephadex A-25, then lyophilized; the titled pentapeptide was obtained.

Owing to the very strong acidity of this crude sample, it was extremely difficult to purify. In RP-HPLC, it could not be retained by the C-18 column. The amino acid analysis demonstrated that the ratio of amino acids of this sample was Asp:Glu:Gly= 1.0:2.8:0.9. The N-terminal was determined to be Glu- by DNS-Cl and C-terminal to be -GlyGluOH

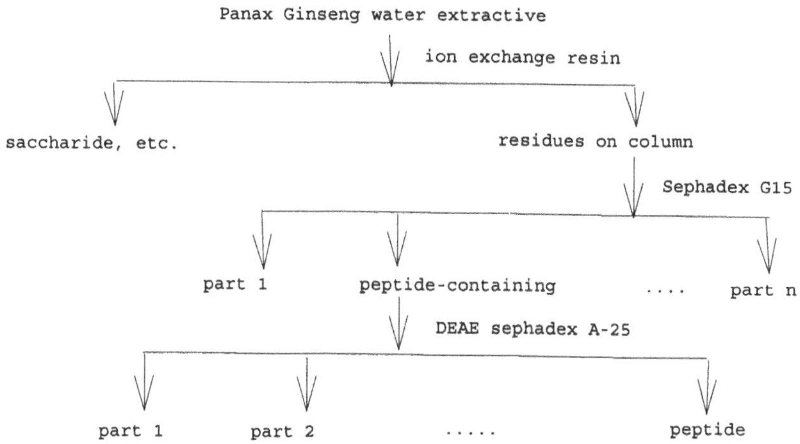

Scheme 1. Isolation procedure

by carboxypeptidase A. It was shown that the content of this pentapeptide was less than one twenty thousandth. The determination of its sequence and bioactivities are in progress.

Acknowledgment

This work was supported by the National Natural Science Foundation of China (No.29132051).

References

1. Yang, L., Ye, Y. H. and Xing, Q. Y., Chinese Chemical Letters, 1 (1990) 51.
2. Yang, L., Ye, Y. H. and Xing, Q. Y., In Peptide: Biology and Chemistry (Proceedings of the 1990 Chinese Peptide Symposium), Beijing: Science Press, 1990.
3. Ye, Y. H., Long, Y. C. and Xing, Q. Y., Proceedings of '92 International Ginseng Symposium, Changchun, China, 1992, p. 98.
4. Xing, Q. Y., Ye, Y. H. and Yang, L., Proceedings of the 6th International Ginseng Symposium, Seoul, Korea, 1993, p. 124.
5. Chen, Z. K., Ye, Y. H. and Xing, Q. Y., Abstracts of '95 International Conference of Ginseng Allie Plants (August 11-14, 1995, Harbin, China), 1995, p. 12.
6. Chen, Z. K., Ye, Y. H. and Xing, Q. Y., Chinese Chemical Letters, 7 (1996) 337.
7. Fan, C. X., Theses of Ph.D., Peking University, Beijing, 1996.

Isolation and synthesis of a group of N-γ-glutamyl oligopeptides from *Panax ginseng*

Chong-Xu Fan, Yun-Hua Ye, Zhi-Kua Chen, Qin Jiang, Liu Yang and Qi-Yi Xing
Department of Chemistry, Peking University, Beijing 100871, China

Introduction

Panax ginseng C. A. Mayer is a widely used herb in traditional Chinese medicine. In the course of our investigation of the water soluble constituents of *ginseng*, some nonprotein amino acids such as γ-aminobutyric acid, β-oxalo-α,β-diamino-propionic acid have been isolated [1, 2]. But ginseng peptides seldom have been studied, possibly because of their occurrence in very low concentrations and difficulties in isolation. In the present work, a group of N-γ-glutamyl oligopeptides, γ-glutamylcysteinylglycine-(2→2')-disulfide (oxidized glutathione, GSSG) **1**, N,N'-bis-γ-glutamylcystinyl glycinamide **2**, γ-glutamylglycinyl-cysteine-(3→3')-disulfide, an isomer of oxidized glutathione (i-GSSG) **3**, N,N'-bis-γ-glutamylcystinyl monoglycine **4**, N-γ-glutamylcystinylglycine **5**, and N^{α}-γ-glutamylarginine **6**, were isolated for the first time from *Panax ginseng*. Compound **3** has never been reported previously. In order to further study their properties and to confirm their structures, we synthesized peptides **2**, **3**, **4**, **5** and **6** by solution method.

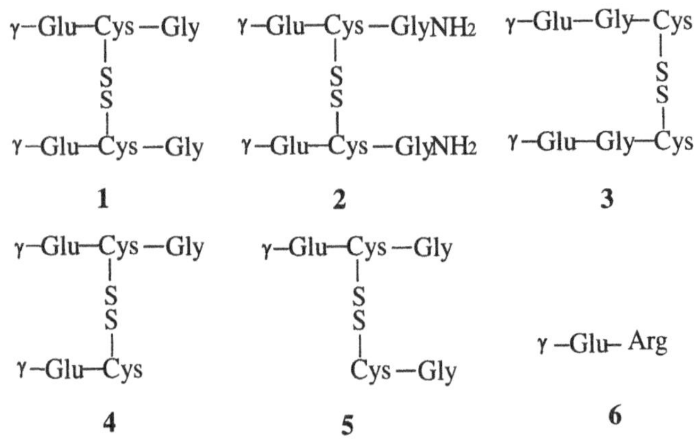

Fig. 1. The structures of the peptides isolated from Panax ginseng.

Results and Discussion

Isolation of the peptides

The powder of ginseng root was extracted with 50% aqueous methanol at -4~0°C. Then, the filtrate was applied to a 001×8 cation-exchange column of H^+ form flushing with water to remove the non-charged substances such as sugar, saponins etc. The total fraction of amino acids, peptides and proteins was eluted with 2M $NH_3 H_2O$. An aqueous solution of total fraction was subsequently subjected to a Dowex 1×8 anion-exchange column of $HCOO^-$ form. Neutral and basic compounds remained in the effluent and the acidic part was eluted with 0.3M HCOOH. The eluate was lyophilized.

The acidic part was subjected to gel filtration on a Sephadex G-15 column and many fractions were obtained. Peptides **1**, **2**, **3**, **4** and **5** were purified by HPLC from fractions 1 and 2.

The neutral and basic part was chromatographed on a Dowex 50W×2 cation-exchange column eluting with a linear gradient of buffer consisting of 0.75M HCOOH +0.2M NH_4OH, pH 3.1 and 1.47M CH_3COOH +1.0M NH_4OH, pH 5.0. Each fraction was subjected to gel filtration of Sephadex G-15 after removing salts with ion-exchange resin and finally a pure peptide **6** was obtained from fractions 5 and 6.

Identification of the pure peptides

1. For determinations of amino acid composition, the peptide samples were hydrolyzed in sealed tube in 6N HCl at 110°C for 24h. The hydrolysates were analyzed with a Beckman 121MB amino acid analyzer.
2. N-terminal and C-terminal amino acids of the peptides were determined by dansylation and hydrolysis of carboxypeptidase A or carboxypeptidase Y.
3. The γ-glutamyl linkage was suggested by the fact that no DABTH-glutamic acid was detected after DABITC/PITC Edman degradation.
4. The molecular weight was determined by FAB-MS.
5. The peptides containing cystine were oxidized with performic acid or reduced with 2-mercaptoethanol to form corresponding sulfonic acids or mercapto derivatives which were analyzed by HPLC. Peptides **4** and **5** gave two peaks because of their two asymmetrical chains.

Thus, the structures were established as shown in figure 1.

Synthesis of peptides

The N-protected peptides **2**, **3** and **6** were synthesized by solution method using 3-O-(diethoxyphosphoryloxy)-1,2,3-benzotriazin-4-(3H)-one (DEPBT) [3] as a coupling reagent or mixed anhydride method. Their structures were confirmed by m.p., FAB-MS, elemental analysis, specific rotation etc. The N-protected peptides **2**, **3** were deblocked by CF_3SO_3H or liquid HF to get compounds **2**, **3**; N-protected peptide **6** was deblocked by hydrogenolysis to obtain compound **6**, and their FAB-MS were $611(M+H)^+$, $613(M+H)^+$ and 304 $(M+H)^+$, respectively.

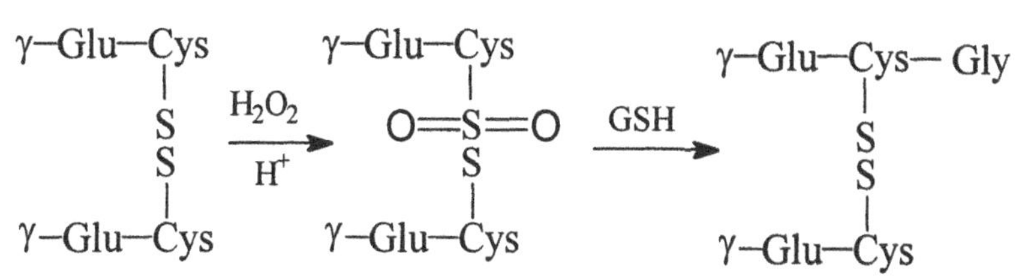

*Scheme 1. The synthesis of peptide **4**.*

The synthesis of asymmetrical disulfides, peptides **4** and **5**, based on glutathione (GSH) thiolysis of the thiolsulfonate derivatives of corresponding peptides (N,N'-γ-glutamylcystine and cystinyl-glycine) is shown in Scheme 1. This method is similar to that of the successful synthesis of cysteine-glutathione mixed disulfide by B. Eriksson [4]. The final products were separated from contaminating compounds by ion-exchange chromatography or HPLC.

Acknowledgments

This work was supported by the National Science Foundation of China (No. 29132051) and the Doctoral Program Foundation of Institution of Higher Education.

References

1. Yang, L., Ye, Y.H., Yuan, H.S. and Xing, Q.Y., Chinese Science Bulletin, 36 (1991) 513.
2. Long, Y.C., Ye, Y.H. and Xing, Q.Y., Int. J. Peptide Protein Res., 47 (1996) 42.
3. Fan, C.X., Hao X.L. and Ye, Y.H., Synth. Commun., 26 (1996) 1455.
4. Eriksson, B. and Eriksson, S.A., Acta Chem. Scand., 21 (1967) 1304.

Studies on the venom of the marine snail *Conus betulinus*

Kai-Hua Wei, Ke-Ping Hu, Lin Yu, Ming-Nai Zhong, Ji-Sheng Chen and Shou-Lan Zhang
Institute of Pharmaceutical Chemistry, Beijing 102205, China

Introduction

Marine snails of the genus Conus are known to be venomous. It is clear that the venom of around 500 species of Conus will be a rich source of biologically active peptides [1]. Studies on the venoms of more than 50 species have been reported, but the chemical and pharmacological properties of the venom of *C. betulinus* have not been examined at all. In the present work, the isolation and toxicology of a number of components from the venom of *C. betulinus* are described.

Results and Discussion

About 10kg of live specimens of *C. betulinus* were collected in Sanya, Hainan province. The shells were 30-150 mm in length. The venom apparatus was composed of venom bulk (5mm in diameter and 20mm in length) and duct (0.5mm in diameter and 100-600 mm in length). 1g of crude lyophilized venom was obtained.

400mg of the peptide fraction from gel filtration (Sephadex G-25, 800× 16 mm x mm, 0.1M NH_4OAc, PH7.0, 20.6 ml/hr) was run on reversed phase HPLC (Hewlett Packard 1050, VWD, flow rate 2.0ml/min,214nm, buffer A=0.1% TFA in H_2O, buffer B=0.1% TFA in CH_3CN,gradient(B%)=0-60%(40min)) using ECONO-PREP C_{18} column (250×10 mm×mm, 10μ particle size), then rerun at smooth gradients, finally run on Nucleosil C_{18} column (250 ×4.6 mm×mm, 5μ). Though more than 50 components were eluted, some of them were not enough for further investigation.

Each component (0.05-1nmol) from HPLC was assayed. Symptoms (Table 1) could be observed after 15-30min and could last for 24 hours. TOF-MS, FAB-MS results of some components showed that the venom of *C. betulinus* may include different types of conotoxins.

One peptide (Be VIII) with Mol. Wt. 3459(TOF-MS) was sequenced: RRAAGTYC-CNNDSQCLLNNCWGWWGGCHCXCR,H= Hyp. Gly^5,Cys^8,Cys^{15},Cys^{20} were conserved in this toxin and ω−conotoxins. It is unusual that BeVIIIl has the odd Cys framework (-c-c-c-c-c-c-c-). Studies of primary structure on some components are in progress.

Table 1. Activities of HPLC components of the venom of C. betulinus

Retention time (min)	Activities (mice, i.c.)
11.6	NA
15.2	NA
15.5	hypoactive, erecting hair, exaggerated respiration
16.5	sensitive
16.9	circling
19.5	NA
19.7	paralysis, dyspneic respiration
22.4	NA
24.5	NA
25.2	paralysis, dyspnea, rolling, circling
26.0	sensitive
26.4	paralysis, circus movement
26.9	hypoactive
27.2	NA
28.9	sensitive, irritation, attacking
29.3	scratching, erecting hair
30.7	circus movement, shaking, sensitive
31.0	hypoactive
31.5	sensitive
32.0	paralysis, exaggerated respiration, erecting hair
32.4	convulsive, rolling, drowsy
33.0	sensitive
33.6	hypoactive, erecting hair, hyperpnea
33.9	paralysis, hyperpnea
34.3	paralysis, convulsive, erecting tail, rolling, exaggerated respiration, circus movement, death
35.0	sensitive, hyperpnea, hypoactive
35.4	sensitive, irritation, attacking
35.8	circus movement, hyperpnea
36.3	sensitive
37.2	NA
38.2	hyperactivity, excitation

NA=no activity.

References

1. Olivera, B.M., J. Biol. Chem., 266 (1991) 22067.

Synthesis of a radioiodinated photoreactive MAGE-1 peptide derivative and photoaffinity labeling of cell-associated human leukocyte antigen-A1 molecules

Fabienne Anjuere[a], Andreas Layer[a], Jean-Charles Cerottini[a], Catherine Servis[b] and Immanuel F. Luescher[a]

[a]*Ludwig Institute for Cancer Research, Lausanne Branch, University of Lausanne, 1066 Epalinges, Switzerland;*

[b]*Institute of Biochemistry, University of Lausanne, 1066 Epalinges, Switzerland*

MHC molecules bind antigen-derived peptides in an allele-specific manner. The resulting MHC class I-peptide complexes expressed on cell surfaces are recognized by $CD8^+$ cytotoxic T lymphocytes (CTL) by means of their antigen receptor, which results in the activation of these cells [1,2]. The formation of relevant MHC-peptide complexes is therefore of great immunological interest. It has been previously shown that the peptide binding by murine MHC class I molecules can be studied on living cells by photoaffinity labeling with adequate photoreactive derivatives of antigenic peptides [3-5]. This photoaffinity labeling was highly specific for the restricting MHC class I molecules in the absence of detectable labeling of other MHC molecules or other cellular components.A novel strategy for the synthesis of radioiodinated photoreactive peptide derivatives was developed, based on simple conventional chemical procedures. As an example, a radioiodinated photoreactive derivative of the MAGE-1 161-169 peptide (EADPTGHSY) was synthesized. This peptide is derived from the MAGE-1-encoded melanoma tumour antigen and is recognized by tumour-specific CTL in the context of HLA-A1 [6]. A photoreactive derivative of this peptide was synthesized by automated solid-phase peptide synthesis in which His-167 was replaced with N-β-4-azidosalicyloyl-2,3-L-diaminopropionic acid (Dap(ASA)). The photoreactive entity, the ASA group, is amenable to iodination similar to tyrosine [7], which provides simple means to incorporate a radiolabel of high specific activity. However, iodination of this peptide derivative, which contains an ASA group and a tyrosine is prone to yield mixed iodination products. This is particularly disturbing in this case because the tyrosine of the MAGE peptide is critically involved in the peptide binding to HLA-A1 [8].

This difficulty can be avoided by blocking the iodination of tyrosine by modification of its phenol group. This, for example has been accomplished by esterification of the tyrosine phenol group with sulphuric acid [9]. The instability of tyrosine sulphates, especially under acidic conditions, however, limits the usefulness of this tyrosine protection. We therefore examined whether tyrosine phosphates could be used instead. Due to the high stability of tyrosine phosphates, the synthesis and deprotection of tyrosine phosphorylated peptides is much less problematic. In addition, due to the growing interest of tyrosine phosphorylated peptides for biological studies, the synthesis of tyrosine phosphorylated peptides has been well established [10]. We found that phosphorylated side chains are inert to oxidative iodination and that this tyrosine protection can be readily removed after the iodination by alkaline phosphatase.

The strategy to combine conventional peptide synthesis to prepare photoreactive peptide derivatives and transient tyrosine phosphorylation for directing radioiodination is a simple and effective way to prepare well-defined photoreactive peptide derivatives of high specific radioactivity. The thus prepared MAGE-1 peptide derivative EADPTGDap(IASA)SY efficiently and selectively photoaffinity labeled cell-associated HLA-A1, but none of the other HLA molecules tested.

References

1. Townsend, A., and Bodmer, H. Annu. Rev. Immunol. 7 (1989) 601.
2. Romero, P., Casanova, J.-L., Cerottini, J.-C., Maryanski, J.L., and Luescher, I.F., J. Exp. Med. 77 (1993) 1247.
3. Luescher, I.F., Lopez, J.A., Malissen, B., and Cerottini, J.-C. J. Immunol. 148 (1992) 1003.
4. Luescher, I.F., Romero, P., Cerottini, J.-C., and Maryanski, J.L. Nature 351 (1991) 72.
5. Reyes, V.E., and Humphreys, R.E., Mol. Immunol. 28 (1991) 341.
6. Traversari, C., van der Bruggen, P., Luescher, I.F., Lurquin, C., Chomez, P., Van Pel, A., De Plaen, E., Amar-Costesec, A., and Boon, T., J. Exp. Med. 176 (1992) 1453.
7. Tae, H. Ji, and Inhae, Ji., Anal. Biochem. 121 (1982) 286.
8. Gaugler, B., Van den Eynde, B., van der Bruggen, P., Romero, P., Gaforio, J.J., De Plaen, E., Leth, B., Brasseur, F., and Boon, T.J., Exp. Med. 179 (1994) 921.
9. Powers, S.P., Pinon, D.I., and Miller, L.J. (1988) Int. J. Peptide Protein Res. 31 (1988) 429.
10. Ottinger, E.A., Shekels, L.L., Bernlohr, D.A., and Barany, G., Biochemistry 32 (1993) 4354.

Development of highly potent human parathyroid hormone analogs

Jesse Z. Dong[a], Michael Chorev[b], Chizu Nakamoto[b], Chee-Wai Woon[a], Barry A. Morgan[a], Michael Rosenblatt[b] and Jacques-Pierre Moreau[a]
[a] *Biomeasure Incorporated, 27 Maple St., Milford, MA 01757, U.S.A.*
[b]*Beth Israel Hospital, Harvard University Medical School, 330 Brookline Ave., Boston, MA 02215, U.S.A.*

Introduction

There is growing evidence that human parathyroid hormone, hPTH(1-84), and its hPTH (1-34) fragment are promising therapeutic agents for treating osteoporosis in that they can stimulate production of trabecular bone in osteoporotic postmenopausal women [1]. Unlike most current therapeutic agents (i.e., estrogen, bisphosphonates, and calcitonin) which only prevent further bone loss, daily injection of hPTH(1-84) or hPTH(1-34) increases bone mass in animals and humans. The unique bone anabolic effect of hPTH has led us to search for more potent hPTH(1-34) analogs.

In aqueous solution the major structural features of hPTH(1-34) are two amphiphilic α-helices located at the N- and C-termini connected by a short length of a structurally undefined hinge region [2] (Fig. 1). Our research is mainly focused on modification of residues in the hydrophobic face of the N-terminal amphiphilic α-helix as well as in the hinge area.

Results and Discussion

PTH(1-34) analogs were synthesized by the rapid Boc-chemistry SPPS strategy developed by Kent et al. [3] and tested in human osteosarcoma cells for stimulating the production of cAMP [4].

It is known that the amphiphilic α-helix has membrane searching and binding properties [5]. There is also evidence implicating an interaction of the N-terminus of PTH with the transmembrane region of the PTH receptor [6]. From such considerations, we hypothesized

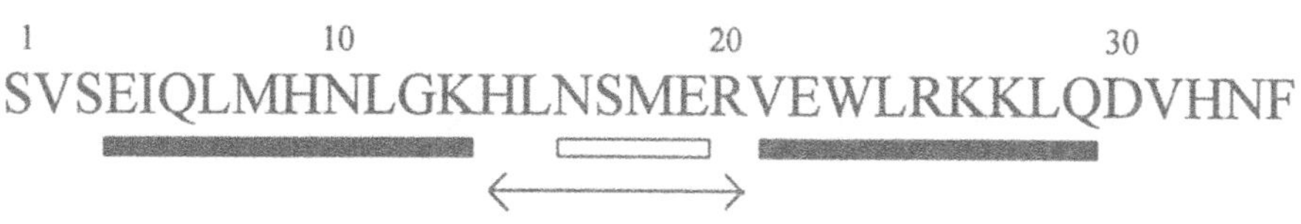

Fig. 1. Structural elements of hPTH(1-34). The regions delineated by solid bars are amphiphilic α-helical domains. The region of the reverse turn is indicated by the open bar. The two-headed arrow denotes the hinge sequence.

that the N-terminal amphiphilic α-helix of PTH could have a membrane binding function which modulates efficient receptor-hormone interaction. Based on this hypothesis, hPTH(1-34) analogs bearing cyclohexylalanine (Cha) in the hydrophobic face of the N-terminal amphiphilic α-helix were designed. The bulky side-chain of Cha increases the size and the hydrophobicity of the hydrophobic face, improving lipid affinity of the peptide.

Replacement of the leucine residue at position 7 in the hydrophobic face of the N-terminal amphiphilic α-helix with Cha gave [Cha^7]hPTH(1-34)NH_2, which is about 2-fold more potent than hPTH(1-34) in stimulating cAMP production (Table 1). A similar improvement in potency (2-fold) was also observed when Cha was substituted for leucine 11. Simultaneously incorporating Cha into these two positions yielded a highly potent analog (peptide #3). This analog has an EC_{50} at 0.6 nM and is 6.7-fold more active than its parent peptide, hPTH(1-34), in cAMP assay. Although the higher membrane affinity of the peptides could contribute to the high potency, some other factors can not be ruled out. The large hydrophobic surface constructed by Cha residues could be favored for the receptor recognition of the hormone. It is also possible that the enlarged surface could facilitate the intramolecular interaction with the hydrophobic face of the C-terminal amphiphilic α-helix [7].

In another series, the hinge region between the two α-helical domains was modified. Substitution of aminoisobutyric acid (Aib) for asparagine 16 or glutamic acid 19 in this region also yielded highly potent analogs, peptides #4 and #5. These analogs are about 5 to 7-fold more potent than hPTH(1-34) in stimulating the production of cAMP. It is possible that Aib at positions 16 and 19 could restrict conformational freedom in this unordered region and increase the population of bioactive conformation in solution[8]. Combining Aib^{16} and Aib^{19} substitutions did not yield a more active analog (peptide #6). However, the $Cha^{7,11}$ Aib^{19} hybrid (peptide #8) with an EC_{50} at 0.5 nM is the most potent analog among these peptides.

In conclusion, several highly potent hPTH(1-34) analogs have been designed and synthesized, which represent a new class of hPTH compounds. These analogs provide new insights on the roles of hydrophobicity in the N-terminal amphiphilic α-helix and the structural flexibility at the hinge region.

Table 1. Biological activity in production of cAMP in human osteosarcoma cells

Peptide	cAMP(EC_{50}) (nM)	Relative Activity to hPTH(1-34)	Sequence
#1	2.3	1.7	[Cha^7]hPTH(1-34)NH_2
#2	2.0	2.0	[Cha^{11}]hPTH(1-34)NH_2
#3	0.6	6.7	[$Cha^{7,11}$]hPTH(1-34)NH_2
#4	0.7	5.7	[Aib^{16}]hPTH(1-34)NH_2
#5	0.6	6.7	[Aib^{19}]hPTH(1-34)NH_2
#6	0.6	6.7	[$Aib^{16,19}$]hPTH(1-34)NH_2
#7	1.1	3.6	[$Cha^{7,11}$,Aib^{16}]hPTH(1-34)NH_2
#8	0.5	8.0	[$Cha^{7,11}$,Aib^{19}]hPTH(1-34)NH_2
hPTH(1-34)	4.0		

References

1. Whitfield, J.F. and Morley, P., TiPS, 16 (1996) 382.
2. Barden, J.A. and Kemp, B.E., Biochemistry, 32 (1993) 7126.
3. Schnolzer, M., Alewood, P., Jones, A., Alewood, D. and Kent, S., Int. J. Peptide Protein Res., 40 (1992) 180.
4. Nakamoto C., Biochem., 34 (1995) 10546.
5. Eisenberg, D., Ann. Rev. Biochem., 53 (1984) 595.
6. Gardella, T.J., Endocrinology, 135 (1994) 1186.
7. Gardella, T.J., J. Biol. Chem., 270 (1995) 6584.
8. Aleman, C., Biopolymers, 34 (1994) 841.

CCK-B receptor activation in human lymphoblastic Jurkat cells results in AP-1 responsive genes activation

D. Gagne, N. Bernad, C. Oiry, E. Cottin, M.-F. Lignon, J. Azay and J. Martinez
Laboratoire des Amino Acides, Peptides et Proteines (LAPP), ESA 5075, CNRS, Universites de Montpellier I et Montpellier II, Faculté de Pharmacie, 15 Av. C. Flahault, F-34060 Montpellier, France

Introduction

1. Extracellular signals regulate many aspects of the growth and differentiation of animal cells. For example, receptors tyrosine kinases (RTK) and G-protein-coupled receptors (GPCR) which are among the various receptors located at the cell membrane are involved in the transfer of regulatory information from the cell surface to the nucleus. Once these receptors are stimulated, mitogen-activated protein kinases (MAPKs), also known as extracellular signal-regulated kinases (ERKs), can be stimulated and thus phosphorylate and regulate key intracellular enzymes and transcription factors involved in the control of cell proliferation as for example Fos and Jun [1].
2. The cell growth is a biological complex response involving several regulatory pathways but in studies relative to gastrointestinal tissues and neoplastic cells, in which gastrointestinal peptides controlled cell proliferation, one can suggest, at least in part, the involvement of their specific receptors since specific antagonists inhibited the agonistic effect. Moreover specific receptors for gastrointestinal peptides have been detected in gastrointestinal tissues and neoplastic cells [2]. Gastrin, cholecystokinin (CCK) and CCK related peptides comprise a neuropeptide family characterized by an identical carboxy-terminal pentapeptide, a critical domain for receptor binding and expression of the biological activity. Gastrin and CCK bind to at least two different receptor subtypes. The peripheral CCK-A receptors are mainly found in the pancreas, gallbladder, ileum and various brain nuclei. In contrast, the CCK-B/gastrin receptors are predominantly found in the cerebral cortex and the gastrointestinal tract. We previously reported on the presence of CCK-B/gastrin receptors on Jurkat cells, a human T lymphoblastic cell line [3, 4]. In addition, we recently cloned and sequenced cDNA encoding the CCK-B/gastrin receptor from such cells.
3. In the present study, we have investigated the trophic effect of CCK-8 in Jurkat cells by analyzing its effect on biological responses known to be regulated by the trans-activator proteins Fos and Jun.

Results and Discussion

We have studied the effect of CCK-8 on AP-1 regulated genes since activation of nuclear oncogenes like Fos and Jun is believed to be involved in events directly related to growth and differentiation of cells [5]. We have used transfected Jurkat cells with a reporter plasmid (p(TRE)3-tk-Luc) in which the reporter protein expression (firefly luciferase, Luc) is enhanced, following Fos and Jun activation [6]. Under these experimental conditions, CCK-8 was able to stimulate luciferase activity in a dose-dependent manner, the maximum induction occurring at 100-200 nM CCK-8. This induction was inhibited by a selective CCK-B receptor antagonist, suggesting the involvement of the CCK-B receptor in this phenomenon. Transient cotransfection of cells free of CCK receptors (COS and CV1 cells) with the plasmid p(TRE)3-tk-Luc and the plasmid encoding the CCK-B receptor from Jurkat cells clearly confirmed these results. As a control, CCK-8 was unable to induce luciferase expression in cells free of CCK-B receptors transiently cotransfected with the plasmid p(TRE)3-tk-Luc and the plasmid encoding the antisense CCK-B receptor of Jurkat cells, or only transfected with the plasmid p(TRE)3-tk-Luc. Quantitatively, 10nM CCK-8 induced luciferase activity to the same extent as 1 nM PMA (or TPA). This corresponds to the effect induced by EGF in breast cancer cells.

These results suggested that 1, the CCK-B receptor, once activated by CCK-8 can contribute to modulate expression of AP-1 regulated genes. To further investigate the mechanism of action, Jurkat cells were transiently transfected with the plasmid p(TRE)3-tk-Luc, and the effect of CCK-8 was studied in the absence and in the presence of two serine/threonine-specific protein phosphatases-1 and -2A (PP-1 and PP-2A) inhibitors (okadaic acid and calyculin A). Using such a reporter plasmid, okadaic acid and calyculin A were shown to drastically increase the PMA (or TPA) luciferase inducibility. In contrast, under the same experimental conditions, okadaic acid and calyculin clearly abolished the CCK-8 effect. These results differentiate CCK-8 from PMA (or TPA) to activate expression of an AP-1 regulated gene. Such a result is in agreement with a recent study showing that 2 CCK may activate MAP kinase in a protein kinase C (PKC) independent manner [7]. This argues in favor that the signal transduction pathway activated by the CCK-B receptor to stimulate MAP kinase is different from that used by other seven-transmembrane domain receptors. Taken all together, these results suggest that 3 activation of the CCK-B receptor by CCK-8 results in the expression of genes regulated by Fos and Jun. We thus conclude that activation of the CCK-B receptor induces a trophic effect on Jurkat cells.

References

1. Cobb, M.H. and Goldsmith, E.J., J. Biol. Chem. 270 (1995) 14843.
2. Silen, M.L. and Gardner, J.D., Trends Endocrin. Met. 4 (1993) 131.
3. Lignon, M.-F., Bernad, N. and Martinez, J. Mol. Pharmacol., 39 (1991) 615.
4. Lignon, M.-F., Bernad, N. and Martinez, J., Eur., J. Pharmacol., 245 (1993) 241.
5. Edwards, D.R., Trends Pharmacol. Sci., 15 (1994) 239.
6. Astruc, M.E., Chabret, C., Bali, P., Gagne, D. and Pons, M., Endocrinology, 136 (1995) 824.
7. Sufferlein, T., Withers, D.J. Broad, S., Herget, T., Walsh, J.H. and Rozengurt, E., Cell Growth & Differentiation, 6 (1995) 383.

Enhancement of MAPK activity in rat brain following AVP(4-8) administration

Li-Ya Qiao and Yu-Cang Du
Shanghai Institute of Biochemistry, Chinese Academy of Sciences, Shanghai 200031, China

Introduction

Mitogen-activated protein kinase (MAPK) can be activated by a variety of extracellular stimuli mediated by respective receptors. It has been reported that arginine-vasopressin fragment(AVP(4-8)) can specifically bind to rat hippocampus with high affinity [1], stimulate the accumulation of second messenger--IP3 [2], and induce the expressions of Fos, nerve growth factor (NGF) and brain-derived growth factor (BDNF) genes [3].The enhance- ment of gene expressions was thought to be a crucial link in the action mechanism of memory-enhancing peptide. Currently, we have shown the involvement of a PTX-sensitive G protein in the AVP(4-8) signaling pathway, but whether a MAPK is also included in this G_0 pathway still remains to be investigated.

Results and Discussion

The brain tissues of Wistar rats were removed and homogenized for a designated time following AVP(4-8) administration(s.c.). The cytosolic extracts obtained by ultra centrifugation were fractionated by MONO-Q anion-exchange chromatography. Fractions were collected and assayed for MAPK activity by phosphorylation of its specific substrate MBP [4]. The protein kinase activity concentrated in the main fraction p indicated that AVP (4-8) significantly enhanced MAPK activity in hippocampus (Fig.1). Furthermore, by using immunoblot assay with anti-ERKI or anti-ERKII polyclonal antibody, the kinase in fraction p was ascertained to be p44ERK (Fig.2). In accordance with immunoblotting result, this MAPK activity, as shown in Fig. 3, is insensitive to PKI, an adenosine 3',5'-cyclic monophosphate-dependent protein kinase inhibitor.

The maximum of MAPK activity stimulated by AVP(4-8) was found to be 3-fold vs control ($P<0.01$) at 2-h in the hippocampus (Fig.4) or 2-fold vs control ($P<0.001$) at 1-h in the cortex (Fig.5) after administration(s.c.) of AVP(4-8). However, immunoblotting assay indicated that MAPK protein level had no detectable difference between the administration groups and control (Fig. 6). These results suggested that the increase of p44ERK activity stimulated by AVP(4-8) was through a short-period activation process caused by protein phosphorylation but not by protein expression.

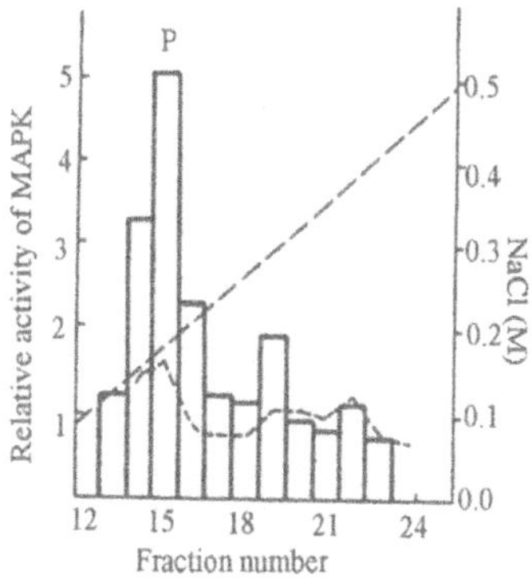

Fig. 1. Fractionation of protein kinase activities on MONO-Q chromatography. Column, AVP(4-8)-treated group; Curve (...), control group.

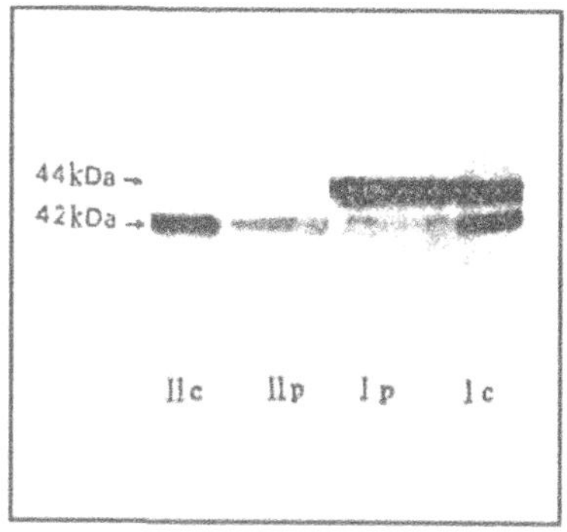

Fig. 2. Immunoblotting with anti-ERKII(IIc & IIp) and anti-ERKI(Ic & Ip) polyclonal antibodies. p, fraction P in Fig.1 from treated group; c, control ample before chromatography.

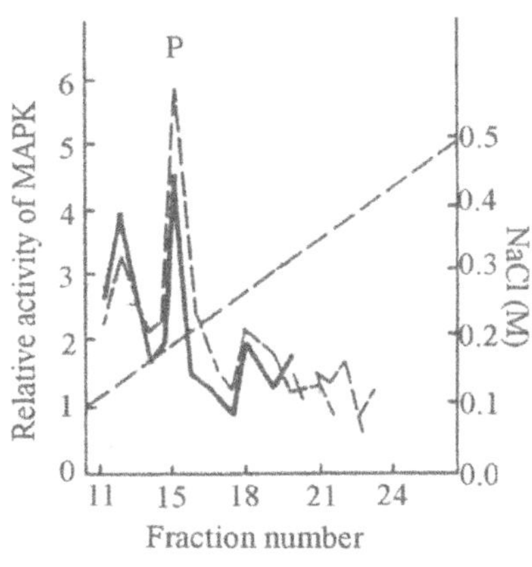

Fig. 3. Profiles of the kinase activity on MONO-Q chromatography in the presence(dotted line) or absence (solid line) of PKI.

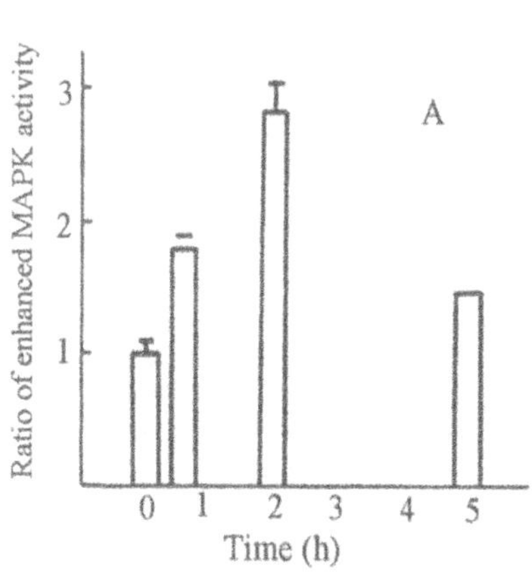

*Fig. 4. Time course of MAPK activity changes in rat hippocampi following the administration of AVP(4-8) (n=5).**, P<0.01; *, P<0.05 vs. control (0).*

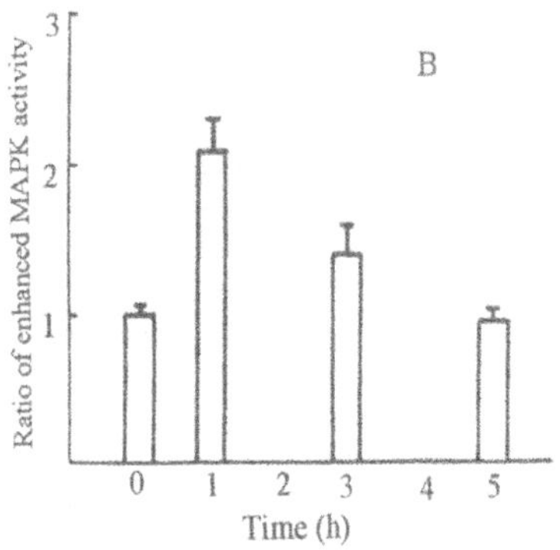

*Fig. 5. Time course of MAPK activity changes in rat cortices following the administration of AVP(4-8)(n=3). ***, P<0.001 vs. control (0).*

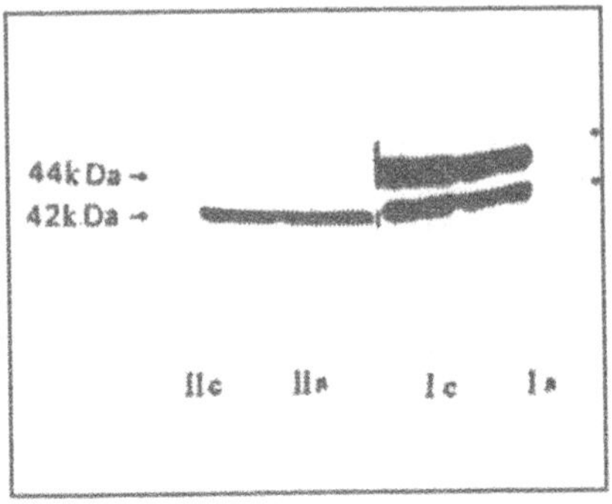

Fig. 6. Immunoblotting analysis of MAPK protein with anti-ERKII (IIc &IIs) and anti-ERKI (Ic & Is) polyclonal antibodies. hippocampal preparation from AVP(4-8)-treated group (s) and control group (c).

Acknowledgment

This work was supported by the National Natural Science Foundation of China, No. 39570157 and the State Key Program No. P8514. AVP(4-8) was synthesized and purified by Jin-Huan Shen and Wen-Yu Wu.

References

1. Du, Y.C., Wu, J.H., Jiang, X.M. and Gu, Y.J., Peptides, 15 (1994) 1273.
2. Gu, B.X. and Du, Y.C., Acta Biochem. Biophys. Sin., 24 (1992) 287.
3. Gu, B.X. and Du, Y.C., Acta Biochem. Biophys. Sin., 23 (1991) 537.
4. Kribben, A., Wieber, E.D., Li, X., Putten, V.V., Granot, Y., Schrier, R.W. and Nemenoff, R.A., Am. J. Physiol., 265 (1991) 939.

Involvement of a G-protein coupled receptor (GPCR) in signal transduction induced by arginine-vasopressin(4-8)

Li-Ya Qiao and Yu-Cang Du
Shanghai Institute of Biochemistry, Chinese Academy of Sciences, Shanghai 200031, China

Introduction

Arginine-vasopressin fragment AVP(4-8) had a high affinity receptor in rat hippocampus [1, 2], and a remarkable induced long-term potentiation (LTP) of synaptic transmission via a N-methyl-D-aspartic acid-receptor (NMDA-R) plus mechanism [3]. It was reported that a pertussis toxin (PTX)-sensitive G protein was involved in tetanus-induced LTP in rat hippocampus [4]. In order to study the receptor that may be involved in LTP process stimulated by AVP (4-8) and the signaling pathway, changes of Mitogen-activated protein kinase (MAPK) activity and Ca^{2+}/CaM-dependent protein kinarseII (CaMKII) autophosphorylation influenced by ZDC(C)PR, PTX, etc. were estimated and their significance was discussed.

Results and Discussion

Wistar rat hippocampi were cut into 300 μm slices and incubated at 37 °C in a humidified atmosphere of 95% O2-5% CO2. Effects of different drugs on MAPK activity and CaMKII autophosphorylation were tested. It was found that the increase of MAPK activity and CaMKII autophosphorylation induced by AVP(4-8) was blocked by ZDC(C)PR, an antagonist of AVP(4-8) (Fig.1), and completely inhibited by PTX, a selective blocker of G_0 protein (Fig.2). These facts indicated that a PTX-sensitive GPCR was involved in AVP(4-8)-induced signaling pathway. Acute treatment with TPA, an agonist of PKC, significantly elicited an increase of MAPK activity, but did not have detectable influence on the activity enhanced by AVP(4-8). PMB, an inhibitor of PKC, thoroughly inhibited the increase of MAPK activity (Fig.3). It was suggested that AVP(4-8)-induced activation of MAPK was mediated by PKC.

In brief, our data clearly demonstrated the existence of a PTX-sensitive G protein, which mediated both mitogenic signaling pathway and CaMKII-LTP induced by AVP(4-8) and the receptor of AVP(4-8) in rat hippocampus should be coupled to G_0 instead of Gi (activating MAPK not through PKC)[5] or Gq (PTX-insensitive) protein[6].

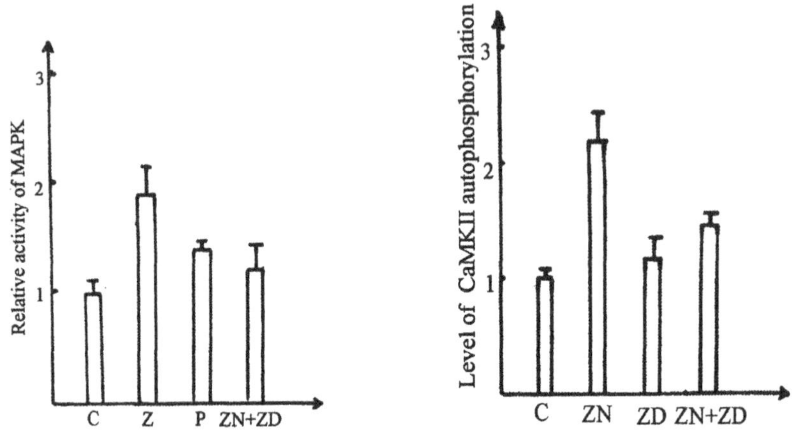

*Fig. 1. Effect of ZDC(C)PR, an antagonist of AVP(4-8) on MAPK activity (A, n=3) and CaMKII autophosphorylation (B, n=3) in rat hippocampal slices pretreated with AVP(4-8). c, control; ZN, AVP(4-8); ZD, ZDC(C)PR; ZN+ZD, AVP(4-8) plus ZDC(C)PR. **, P<0.01, *, P<0.05 vs.control. a, P<0.05 vs.ZN.*

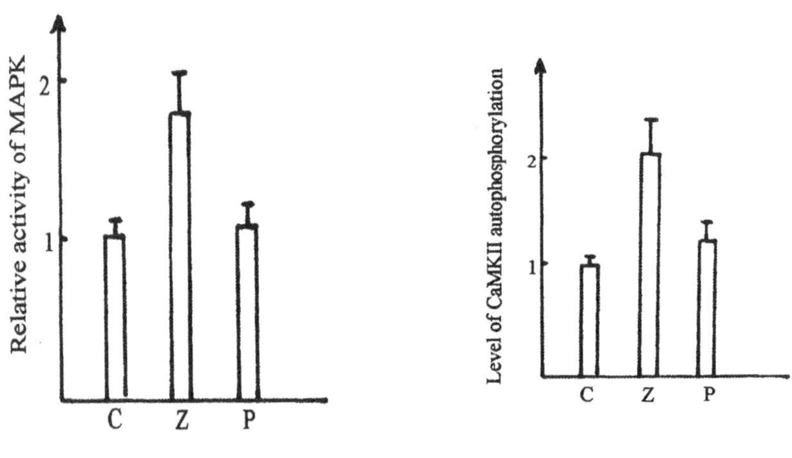

*Fig. 2. Effect of PTX on MAPK activity (A, n=5) and CaMKII autophosphorylation (B, n=5) in hippocampal slices incubated with AVP(4-8). C, control; Z, AVP(4-8); P, AVP(4-8) following PTX preincubation. **, P<0.01; *, P<0.05 vs.control. a, P<0.05 vs.Z.*

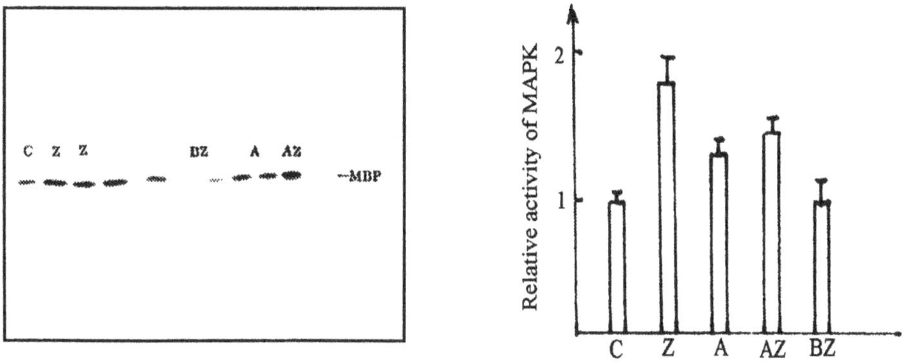

*Fig. 3. Effect of TPA and PMB on MAPK activity in hippocampal slices in the presence of AVP(4-8)(n=5). C, control; Z, AVP(4-8); A, TPA; AZ, AVP(4-8) plus TPA; BZ, AVP(4-8) plus PMB. ***, P<0.001, **, P<0.01; *, P<0.05 vs. control. a, P<0.001 vs. Z.*

Acknowledgment

This work was supported by grants from the State Key Program, National Natural Science Foundation and Shanghai Life Science Center. AVP(4-8) was synthesized and purified by Jin-huan Shen and Wen-yu Wu. TPA and PMB were kindly provided by Prof. Xiu-fang Chen.

References

1. Du, Y.C., Guo, N.N. and Chen, Z.F., Acta. Physiol. Sin., 46 (1994) 435.
2. Du, Y.C., Wu, J.H., Jiang, X.M. and Gu, Y.J., Peptides, 15 (1994) 1273.
3. Rong, X.W., Chen, X.F. and Du, Y.C., Neuroreport, 4 (1993) 1135.
4. Goh, J.W. and Pennefather, P.S., Science, 244 (1989) 980.
5. Koch, W.J., Hawes, B.E., Allen, L.F. and Lefkowitz, R.J., Proc. Natl. Acad. Sci. USA, 91 (1994) 12076.
6. Troppmair, J., Bruder, J.T., Munoz, H., Lloyd, P.A., Kyriakis, J., Baneriee, P., Avruch, J. and Rapp, U.R., J. Biol. Chem., 263 (1993) 7030.

Effect of arginine vasopressin AVP(4-8) on CaMKII autophosphorylation and CaM expression in rat brain

Li-Ya Qiao, Xiu-Fang Chen, Ben-Xian Gu and Yu-Cang Du
Shanghai Institute of Biochemistry, Chinese Academy of Sciences, Shanghai 200031, China

Introduction

The memory-enhancing peptide, AVP(4-8), existing in mammalian brain as a metabolite of arginine vasopressin (AVP), facilitates the short-term and long-term learning and memory of animals. Significant acceleration of phosphorylated growth-associated protein (GAP43) maturation by AVP(4-8) [1] may be functionally involved in synaptic plasticity, axon elongation and long-term potentiation (LTP) [2]. Calmodulin (CaM) is a ubiquitous intracellular calcium receptor which can be released from CaM-binding proteins, such as GAP43 and MARCKS when they are phosphorylated by protein kinase C(PKC) [3]. CaM is responsible for activating more than 20 enzymes [4], among them Ca^{2+}/CaM-dependent protein kinase (CaMKII) is a crucial enzyme for LTP. After Ca^{2+}/CaM-dependent autophosphorylation, CaMKII displays its protein kinase activity by switching the molecule into an "on" state. It has been found that LTP can be promptly induced by AVP(4-8) in rat hippocampual slices, but the process of LTP may not be necessarily related to the activation of NMDA receptor. So it is interesting to observe the changes of effctive CaM level and CaMKII activity in rat brain when the animals are administered with AVP(4-8).

Results and Discussion

The brain tissues of Wistar rats were removed and homogenized at designated times after AVP(4-8) administration(s.c.). The level of CaMKII autophosphorylation in brain cytosolic extracts was determined in the presence of Ca^{2+}. A maximal increase in cerebral cortex stimulated by AVP(4-8) was observed to be 2-fold ($P<0.001$) vs. control at h-1, while 40% increase ($P<0.05$) was found in hippocampus (Fig.1). CaMKII activity repre-sented by its autophosphorylation level appeared to be dependent on the concentrations of Ca^{2+} and CaM (Fig.2). Therefore the activation of CaMKII can be regulated by the release of CaM from phosphorylated GAP43. Effective CaM level and its gene expression were separately analyzed by enzymatic determination [5] and by Northern blot assay. Total RNA in rat hippocampus was extracted and hybridized with CaM or GAPDH cDNA probes. The level of CaM mRNA was densitometrically determined on Northern blots and normalized by GAPDH mRNA, one hour after AVP(4-8) administration. The effective CaM in plasma increased to 3.4-fold ($P<0.05$), while the level of CaM mRNA was not detectably altered. However, the CaM transcription significantly rose to a maximum of 2.5-fold ($P<0.01$) in three hours and the protein level was 5.8-fold ($P<0.001$) vs. control (Fig.3). It means that the increase of CaM may occur at two levels, i.e., delivery from the CaM pool (GAP43) and expression of CaM gene.

Since the enhancement of CaM expression may replenish CaM pool and maintain high activities of CaMKII and/or other CaM-coupled enzymes, the sustained activation of CaMKII would lead to the enhancement of synaptic transmitter release and LTP phenomena. Therefore, the involvement of both CaM and CaMPKII in the peptide-induced LTP and memory formation process was suggested.

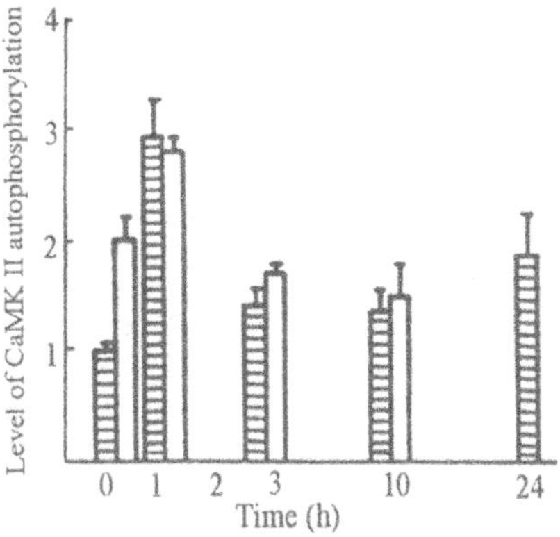

*Fig. 1. Time course of changes of CaMKII autophosphorylation level in hippocampi (blank column, n=3) and cortices (shaded column, n=5) following AVP(4-8) nistration. ***, P<0.001; *, P<0.05 vs. control (0).*

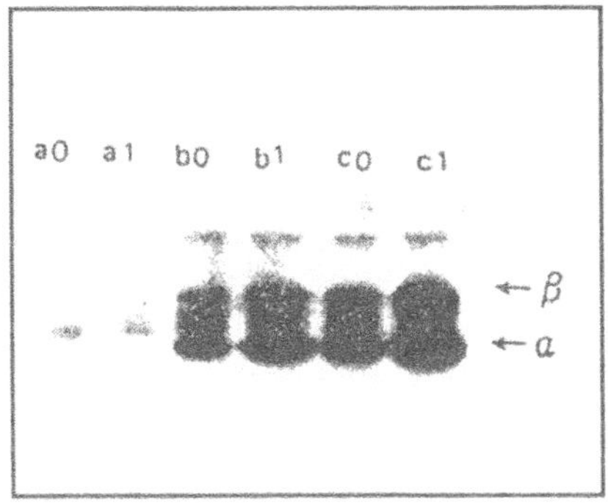

Fig. 2. Effect of Ca^{2+} and CaM in reaction system on autophosphorylation of CaMKII. ***a,*** *in the absence of Ca^{2} and CaM;* ***b,*** *in the presence of Ca^{2+} and* ***c,*** *in the presence of Ca^{2+} and CaM.*

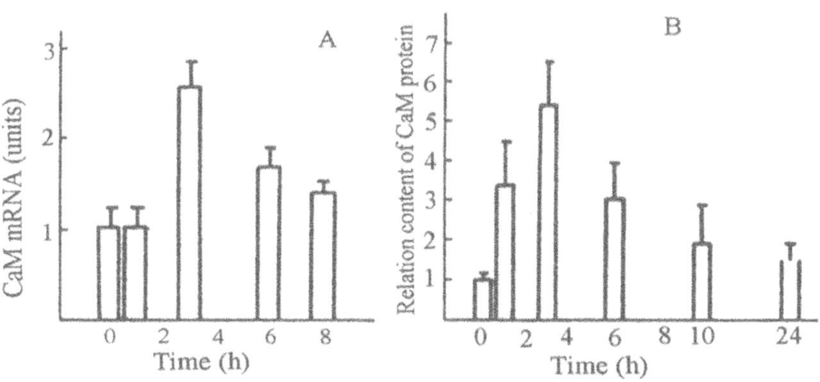

*Fig. 3. Time course of the increases of CaM mRNA level (A, n=3) and CaM protein content (B, n=6) in cytosolic lysate of hippocampi after AVP(4-8) administration(s.c.). ***, P<0.001; **, P<0.01; *, P<0.05 vs. control (0).*

Acknowledgment

This work was supported by grants from the State Key Program, National Natural Science Foundation and Shanghai Life Science Center. AVP(4-8) was synthesized and purified by Jin-Huan Shen and Wen-Yu Wu.

References

1. Chen, X.F., Tang, T., Zhang, J.W., Miao, H.H., Wang, T.X. and Du, Y.C., Mol. Repro. Devel., 35 (1993) 251.
2. Rong, X..W., Chen, X.F. and Du, Y.C., NeuroReport, 4 (1993) 1135.
3. Chen, S.J., Klann, E. and Sweatt, J.D., Soc. Neurosci. Abstr., 19 (1993) 1707.
4. Olwin, B.B. and Storm, D.R., Biochemistry, 24 (1985) 8081.
5. Sharma, R..K. and Wang, J.H., Adv. Cyclic Nucleo. Res., 10 (1979) 187.

Synthesis of phosphopeptides as inhibitors of progesterone production

Qing-Chai Xu, Gui-Shen Lu and Ying Sun
Institute of Materia Medica, Chinese Academy of Medical Sciences and Peking Union Medical College, Beijing 100050, China

Introduction

Many cellular functions are modulated by protein kinases and protein phosphatases which control phosphorylation and dephosphorylation of proteins. Therefore, phosphorylated peptides and proteins may play important roles in regulatory processes [1-2]. Some peptides with anti-progesterone production effect on corpus luteum cells were phosphorylated *in vitro*, and the activities of these phosphopeptides will be further investigated.

Results and Discussion

Tyrosine was reported to have inhibiting activity on hCG-induced progesterone production (hCG-PP) [3]. Further investigation showed that serine and threonine had the same activity as tyrosine. A series of peptides were designed and synthesized and their activities determined [4]. The results showed that some of the synthetic peptides had higher controlling activities on progesterone production than tyrosine. All these peptides contain tyrosine, serine or threonine. The activities of decreasing progesterone production of these peptides might be induced by activating protein phosphatase C and protein kinase C.The mechanism of anti-progesterone production may be related to the phosphorylation of proteins. It is well known that the phosphorylation of proteins on tyrosine, serine or threonine residues plays important roles in regulatory processes.Therefore, it is possible to find compounds with high activity if peptides possessing anti-progesterone production effects are phosphorylated.

Eight active peptides with the structures given in Fig. 1 were phosphorylated. Firstly, the protected peptides with the phosphorylation sites unprotected were synthesized.Then the

GlyTyr(PO_3H_2)AlaLys (I)

GlyTyr(PO_3H_2)NH_2 (II)

GlyTyr(PO_3H_2)$NHNH_2$ (III)

SarSer(PO_3H_2) / SarSer(PO_3H_2) > Lys (IV)

SarThr(PO_3H_2) / SarThr(PO_3H_2) > Lys-OMe (V)

Tyr(PO_3H_2)Lys (VI)

Tyr(PO_3H_2)ThrProArgLys (VII)

TyrThr(PO_3H_2)ProArgLys (VIII)

Fig. 1. The structures of eight active peptides.

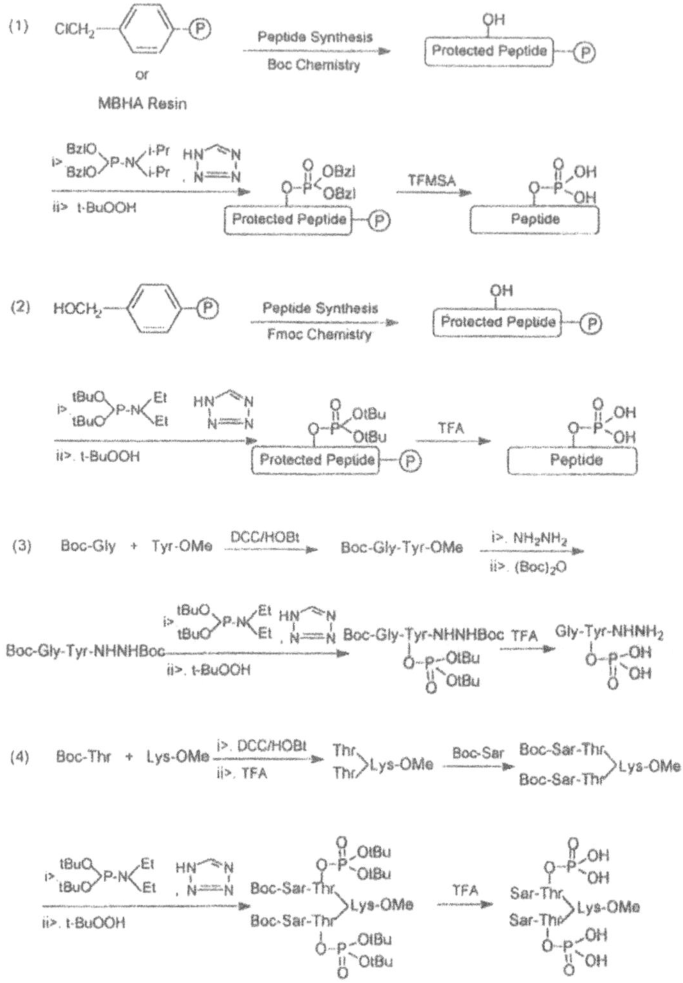

Fig. 2. Synthetic routes of the eight phosphopeptides.

phosphorylation was carried out, and finally the protected groups were cleaved to obtain the target peptides. Four synthetic routes (Fig.2) were adopted [2, 5].

Peptides II, IV,VII and VIII were synthesized by route(1), peptides I and VI were prepared by route (2). Target III and V were prepared by route (3) and (4) respectively.

Crude products were purified by semi-preparative HPLC with 10m-Bondapak TMC18 column (10 x 250 mm). Buffer system: A, 0.1% TFA/H_2O; B, 0.1% TFA/MeOH. All compounds were examined by amino acid analysis and ^{31}P-NMR (Table 1). The bioactivities of these phosphopeptides will be described in another paper.

Table 1. The amino acid analysis and the chemical shifts of ^{31}P-NMR

Peptide	Gly	Ser	Thr	Tyr	Lys	Arg	Pro	Ala	^{31}P-NMR (ppm)
I	1.00			0.99	1.09			1.02	-3.24
II	1.07			1.00					-3.23
III	1.00			1.00					-3.23
IV		2.11			1.00				+7.17
V			2.10		1.00				+3.74
VI				1.00	1.06				-3.17
VII			1.00	1.30	1.00	0.91	1.05		-3.63
VIII			0.87	0.87	1.00	1.00	1.00		+0.16

Acknowledgment

This work was supported by Foundation of Chinese Academy of Medical Sciences 95310017.

References

1. Cohen, P., Nature, 296 (1982) 613.
2. Andrews, D.M., Kitchin, J. and Seale, P.W., Int. J. Peptide Protein Res., 38 (1991) 469.
3. Cheng, Z.P., Reproduction and Contraception, 2 (1982) 23.
4. Wang, D.X.,Wang, N., Lu, G.S. and Xu, Q.C., Acta pharmaceutica Sinica, 29 (1994) 438.
5. De Bont,H.B.A., Liskamp, R.M.J., O'Brain, C.A., Erkelens, C., Veeneman, G.H. and Van Boom, J.H., Int. J. Peptide Protein Res., 33 (1989) 115.

Synthesis of DL-2-amino-3-(1,2-dihydro-2-thio-quinolinyl)propionic acid

Ben-Zhi Qiao, Ai-Xiu Li and Shi-Ying Jia
Department of Applied Chemistry, Taiyuan University of Technology, Taiyuan 030024, China

Introduction

DL-2-amino-3-(1,2-dihydro-2-oxo-quinolinyl) propionic acid (Aoq) have been synthesized [1, 2] and resolved [2]. The physiological activities of DL-, D- and L-Aoq have been reported [2, 3]. In order to study the physiological activities of its sulfur substituted analogue-Atq, ethyl DL-2-acetylamino-3-(1,2-dihydro-2-thio-quinolinyl) propionate (I) as a precursor of Atq should be synthesized via selective thionation of ethyl 2-acetylamino-3-(1,2-dihydro-2-oxo-quinolinyl) propionate (II).

Results and Discussion

The thionation reaction could occur on three functional groups of compoud II, there are two amide groups and an ester group, but they have different features during the thionation procedure. However, thioamides are generally very stable, sometimes even more stable than amide in alkaline medium [4]. It is suggested that thioamides have resonance structures because of large contributions from sp^2 hybrid orbitals of the central carbon and nitrogen atoms. In general, sulfurizing of the carbonyl groups of esters is difficult because of the effection of alkoxy, the reaction occurs only in special cases [5], and is often inferior to that of amides in the competition of thionation [6].

According to the theories of reactions, the carbonyl group site **a** was more difficult to be sulfurized than the amide group of site **b** and site **c**. The thionation of amide belongs to nucleophilic substitution. Site **b** was liable to be sulfurized because of less S_N2 steric hindrance; But the sulfurized product of site **c** was more stable because the conjugative effect reduced the energy of the thioquinolinyl cycle and rendered its structure more stable. So, the adoption of higher temperature, lower concentration and longer reaction time or other intensive measurements can benefit the sulfurizing of site **c**. The expected result was obtained, and the structures of the title compounds have been confirmed.

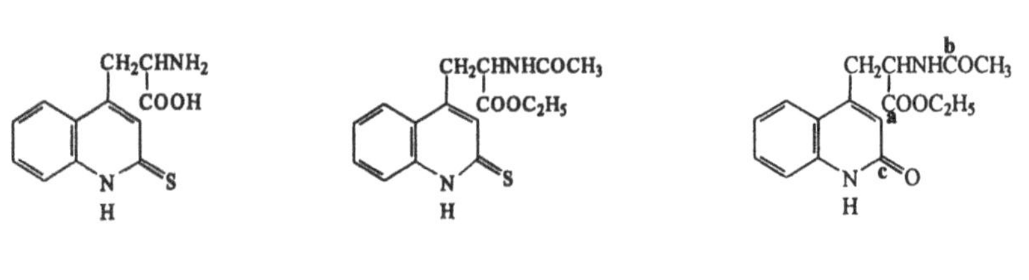

Atq I II

Atq and **I** have not been reported before, the resolution of DL-Atq and the evaluations of physiological activities of these compounds are in progress.

Preparations of Atq and I

1.2g (4 mmol) II was treated with 0.89g (4 mmol) P_2S_5 or 0.81g (2 mmol) p-methoxy-phenylthionophosphine sulphide in benzene and refluxed for 4.0 hr to give 0.51g (1.60 mmol) or 0.42g (1.32 mmol) I respectively. 0.38g (1.20 mmol) I refluxed in 30 ml 2N KOH for 8.5 hr to give 0.23g (0.93 mmol) DL-Atq. (yield 23.15%) m.p. 321~322℃. Anal.Calcd. for $C_{12}H_{12}O_2N_2S$(%): C, 58.05; H, 4.87; N, 11.28; S, 12.91. Found: C, 58.10; H, 4.83; N, 11.25; S, 12.86. IR(KBr) vmax (cm^{-1}) 3435(OH, NH_2), 1690(C=O), 1080(C=S). ^{1}NMR δ(ppm): 4.51, 5.28, 6.93, 7.21. MS(EI) m/e: 248, 204, 174.

Acknowledgment

The authors thank Professor Yun-Hua Ye of Peking University for helpful discussion.

References

1. Uchida, M. and Tabusa, F., Chem. Pharm. Bull., 34 (1996) 4821.
2. Ye, Y.H., Ji, J.G., Qiu, X.S., Liu, S.Y., Zhang, W.Y., Wang, Z.S.and Dai, X.J., Acta Scientiarum Naturalium Universitatis Pekinensis, 26 (1990) 207.
3. Ye, Y.H., Zhang, D.Y., Ji, J.G., Liu, S.Y., Dai, X.J. and Zhang, W.Y., Abstracts of 3rd Sino-French Symposium on the Chemistry of Natural Products, (October 5~7, 1993, Shanghai, China) p. 55.
4. Barton, D. and Ollie, W.D., Comprehensive Organic Chemistry, Vol. 3, p. 440.
5. Kaloustian, M.K. and Khouri, F., Tetrahedron Letters, 22 (1981) 413.
6. Sunderg, R.J., Walters, C.P. and Bloom, J.D., J.Org. Chem., 46 (1981) 3730.

The relationship between hypotensive effect and structure of adrenomedullin

Shu-Li Sheng, Rong Wang, Yu-Hong Guo, Zhi-Juan Ji, Zhi-Wei Zhao and Ren-Ling Zhang
Polypeptide Laboratory, Beijing Xuanwu Hospital, Beijing 100053, China

Introduction

Adrenomedullin (AM) is a novel hypotensive peptide isolated from human pheochromocytoma in 1993. It consists of 52 amino acid residues with a disulfide bond from position 16 to 21. In order to determine the relationship between hypotensive effect and structure of AM, we synthesized seven fragments of human AM (hAM1-15, 16-52, 23-52, 22-42, 26-42, 28-42,. 40-52) and one fragment of rat AM (rAM1-13) by solid-phase method. These fragments purified by HPLC were injected intravenously to rats and the blood pressure(Bp) was monitored.

Results and Discussion

hAM16-52, 23-52, 22-42, 26-42 and 28-42 could decrease the Bp of rats by 7.0-24.5mmHg as soon as they were injected intravenously (Fig.1). This effect lasted 2-10 minutes and the fragments with or without a disulfide bond had similar hypotensive effect (Table1). N-terminal fragments (rAM1-13, hAM1-15) and C-terminal fragment (hAM40-52), however, had no influence on rat Bp. According to our experiment, the molecular structure that decreasing Bp is positioned in the central region of AM. AM28-42 is probably the shortest sequence needed to maintain the hypotensive effect, suggesting that a disulfide bond is not necessary for these effects.

Table 1. The Effect of Fragments of AM on Bp of rats by Intravenous Injection

fragments of AM (n=10)	hypotensive effect (%)	mean depress Bp value(mmHg)
rAM13-52	0	0
hAM1-15	0	0
hAM16-52	100	16.06+7.13
hAM23-52	98	15.76+7.87
hAM22-42	96	15.46+8.28
hAM26-42	106	17.00+7.53
hAM28-42	97	15.51+7.43
hAM40-52	0	0

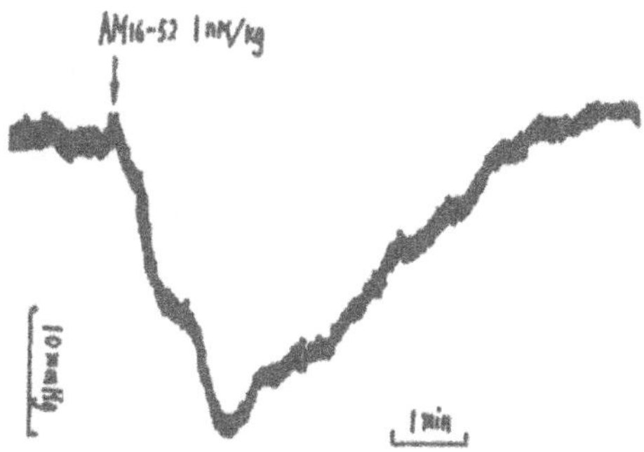

Fig. 1. The Effect of hAM16-52 on Bp of rats by Intravenous Injection.

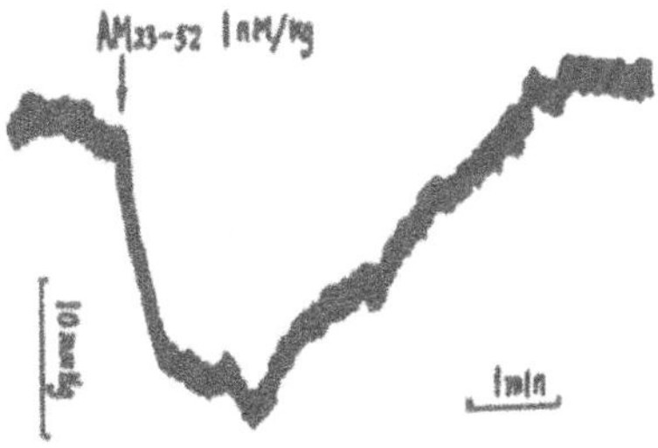

Fig. 2. The Effect of hAM23-52 on Bp of rats by Intravenous Injection.

References

1. Kitamura K., Kangawa K. and Kawamoto M., Biochem. Biophy. Res. Commun., 192 (1993) 553.
2. Skata, J., Shimokubo,T., and Kitamura, K., Biochem. Biophys. Res. Commun., 195 (1993) 921.
3. Tian, Q., Zhao, D., Tan, D.Y., Chinese Journal of Hypertension, 2 (1994) 13.
4. Qi, H.W., Zhao, Y.M. and Liu, Y.Q., Chinese Journal of Diabetes, 4 (1996) 50.

The preliminary study of arrhythmias in rats caused by PAMP

Shu-Li Sheng, Ren-Ling Zhang, Rong Wang, Zhi-Wei Zhao, Yu-Hong Guo and Zhi-Juan Ji

Polypeptide Laboratory, Beijing Xuanwu Hospital, Beijing 100053, China

Introduction

Adrenomedullin(AM) was discovered in 1993 in human pheochrocytoma, PAMP(ProAM 22-41) was considered to be a novel hypotensor. This peptide is widely distributed in many tissues, with the highest concentrations in adrenal medulla and cardiac atria. Because pheochromacytoma patients suffered from arrhythmia during the onset of hypertension, PAMP may be involved in arrhythmia.

Results and Discussion

In the dose-response curve, PAPM and CGRP can decrease Bp, although, CGRP can not induce any abnormality of cardiac rhythm. Sodium nitroprusside at 1mg/kg decreased Bp by 34.50+11.37mm Hg equivalent to the Bp decrease induced by PAMP at 30.0nM, but no arrhythmia was found in the electrocardiogram. A variety of PAMP-induced arrhythmias are shown in Table 1. Table 2 indicates that the Bp decline of all rats with PAPM(30.0nM) is of the same grade regardless of arrhythmia. Table 3 reveals that PAMP(0.1nm) administered simultaneously with NE or E can increase significantly the incidence of arrhythmia.

PAMP is known to be a potent dilatator of blood vessels. The efficacy of BP decrease is similar to that of CGRP, which is the strongest vasodilator currently known.However, PAMP can induce rat arrhythmia. The more PAMP administered, the higher the incidence

Table 1. The number of arrhythmias caused by PAMP and scores of arrhythmias

Group (n)	SB	AV block I	AV block II	AV block III	ME	VT	VF	DT	Score of arrhythmia
NS(40)									5.00 ± 0.00
PAMP	0.1nM(12)	1			1				4.75 ± 0.83**
	1.0nM(12)		1		1				4.75 ± 0.83**
	10.0nM(12)	1		2	8	3			3.76 ± 1.77*
	30.0nM(12)	3	3		6		3	2	2.88 ± 1.87*

* P<0.05 compare with NS ** P<0.001 compare with PAMP(30nM) SB: sinus bradycardia
VF: ventricular fibrillation AV: atrioventricular block DT: death
ME: multiple extrasystole VT: ventricular tachycardia

Table 2. The relationship between arrhythmia caused by PAMP(30.0nM) and blood pressure

Group	Arrhythmia (n=11)	Noemal (n=6)
decrease(mmHg)	29.70± 8.97	32.20± 9.60

Table 3. Effect of interaction of PAMP and NE or E on incidence of rat arrhythmia

Group (n)	A (21)	B (13)	C (12)	D (9)	E # (9)	F* (33)	G**(25)
Chemicals and dose	NE 0.5 μg/kg	E 0.125μg/kg	PAMP 1.0 nM	NE0.5 μg/kg PAMP1.0mM	E0.125 μg/kg PAMP1.0mM		
Incidence of arrhythmia(%)	19.1	30.8	8.3	66.7	77.8#	15.1	20.0

* F is the sum of A and C; ** G is the sum of B and C; P<0.01 compared with F; # P<0.01 compared with G

of arrhythmia. In order to lower the incidence of arrhythmia, we used CGRP and sodium nitroprusside to decrease Bp to about the same level as that induced by PAMP, but no arrhythmia occurred. It can be concluded that PAMP can induce rat arrhythmia which is unrelated to Bp decrease. Table 3 shows that PAMP can enhance the incidence of arrhythmias caused by NE or E. The cause of this enhancing effect is not clear.

References

1. Kitamura, K., Kangawa, K. and Kawamoto, M., Biochem. Biophys. Res. Commun., 192 (1993) 553.
2. Kitamura, K., Sakate, J. and Kanganna, K., Biochem. Biophys. Res. Commun., 192 (1993) 720.

Synthesis and bioactivity of Hir-CGRP chimeric peptides

Liang-You Wang, He-Ping Pan and Zheng-Ying Chen
Institute of Basic Medical Sciences, AMMS, Beijing 100850, China

Introduction

Hirudin (Hir) is a single chain peptide of 65 amino acids isolated from medicinal leech extracts, which is a highly specific α-thrombin inhibitor [1]. Structure-function studies of Hir found that the minimal fragment which retains reasonable anticoagulant activity was Hir53-64 [2]. Hirulog-I, an analog of Hir has higher activity than Hir [3]. Calcitonin gene-related peptide(CGRP) is a pleiotropic neuropeptide of 37 amino acids with a disulfide bond. It has a wide range of biological effects especially in the cardiovascular and cerebral vascular systems [4].

Hir53-64	H-Asn-Gly-Asp-Phe-Glu-Glu-Ile-Pro-Glu-Glu-Tyr-Leu-OH
Hirulog-I	H-(D-Phe)-Pro-Arg-Pro-(Gly)4-Asn-Gly-Asp-Phe-Glu-Glu-Ile-Pro-Glu-Glu- Tyr-Leu-OH
CGRP	H-Ala-Cys-Asp-Thr-Ala-Thr-Cys-Val-Thr-His-Arg-Leu-Ala-Gly-Leu -Leu-Ser-Arg-Ser-Gly-Gly-Val-Val-Lys-Asn-Asn-Phe-Val-Pro-Thr-Asn-Val-Gly-Ser-Lys-Ala-Phe-NH_2

Because both Hir and CGRP are promising drugs for cardiovascular and cerebral vascular diseases, we designed and synthesized the following two chimeric peptides (P1 and P2) to study their biological activities.

P1	Hir53-64-$(Gly)_4$-CGRP
P2	Hirulog-I-$HN(CH_2)_5CO$-CGRP

Results and Discussion

Peptides P1 and P2 were synthesized from N^{α} -Fmoc-amino acids using standard solid phase peptide procedure and HBTU-HOBT-NMM as coupling reagent. Peptides were cleaved from the resin with TMBS/Thioanisole/TFA in the presence of two scavengers m-cresol and EDT, in an ice bath for 1h. All side-chain protecting groups were cleaved at the same time. The disulfide bond was formed by 10% DMSO/TFA in 25°C for 1h. The peptides were purified by G-15 and HPLC. Preliminary pharmacological observations of P1 and P2 were performed. P1 and P2 have the double effects of prolonging thrombin time and reducing rat blood pressure (Fig. 1 and Fig. 2).

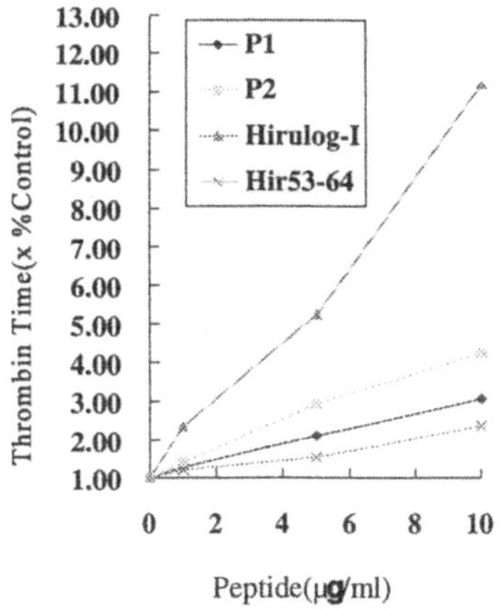

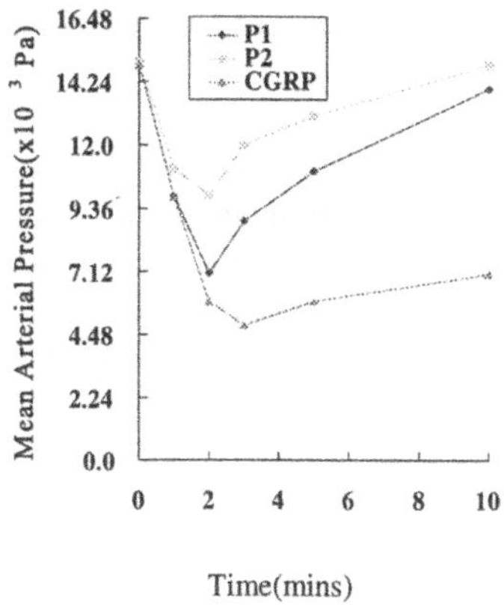

Fig. 1. Anticoagulant activity of P1, P2, Hir 53-64 and Hirulog-I.

Fig. 2. The change in the mean artrial pressure of the rat with P1, P2 and CGRP (10 nmol/Kg).

References

1. Jakubowski, J.A. and Maraganore, J.M., Blood, 75 (1990) 399.
2. Maraganore, J.M., Chao, B., Joseph, M.L., Jablonski, J. and Ramachandran, K.L., J. Biol.Chem., 4 (1989) 8692.
3. Maraganore, J.M., Bourdon, P., Jablonski, J., Ramachandran, K.L. and Fenton, J.W., Biochemistry, 29 (1990) 7095.
4. Marshalls, I., Al-Kazwini, S.J., Robert, P.M., Shepperson, N.B., Adams, M. and Craig, R.K., Eur. J. Pharmacal., 123 (1986) 207.

The cardiovascular effects of YIGSR and RGD containing peptides

Zi-Lin Chen[a], Li-Ping Tu[a], Karl Specht[a], Li Zhao[a], Ming Zhao[b], Dilinur Sabitouf[b], Chao Wang[b] and Shi-Qi Peng[b]

[a] *Jiangxi-OAI Gemeinsames Institut, Nanchang 330047, China*

[b] *School of Pharmaceutical Sciences, Beijing Medical University, Beijing 100083, China*

Introduction

The process of cancer cell invasion through basement membrane is a critical step in the development of metastatic disease. Glycoprotein laminin and fibrinogen appear to be involved in the mediation of this invasion process via its binding to high affinity-receptors. This means there is some relationship between metastasis and thrombosis. In the present study RGDS(1), RGDV(2), RGDF(3), YIGSR(4) were selected to constitute YIGSRRGDS (5), YIGSRRGDV(6), YIGSRRGDF(7) and their antiaggregation effects were observed.

Results and Discussion

Peptides 1-7 were synthesized with protected L-amino acids by a solution method (Fig.1). Their antiaggregation effects (Table 1-2) were inductor and sequence-dependent.

Table 1. Effect of the Peptides in the Maximal Rate of Platelet Aggregation Induced by ADP

		Rate of Maximal Aggregation (%)	
Comp.	ADP(mol/l)	Without peptide	With peptide
1	5×10^{-6}	21.74±5.29	14.80±2.77
2	5×10^{-5}	13.50±2.92	7.53±2.43*
3	5×10^{-5}	38.11±4.91	33.64±7.15
4	5×10^{-6}	28.24±3.14	27.01±2.27
5	5×10^{-6}	23.43±2.10	22.24±2.11
6	5×10^{-6}	25.77±1.38	18.50±2.45
7	5×10^{-6}	25.17±1.38	24.19±3.13

Compare to without peptide *$P<0.05$; n=4.

R+{[D+S(V,F)]} → RGDS(**1**), RGDV, RGDF(**3**)

{**4** +**1-3**} → YIGSRRGDS(**5**), YIGSRRGDV(**6**), YIGSRRGDF(**7**)

Y+{I+[G+(S+R)]} → YIGSR(**4**)

Fig. 1. The Synthetic Route.

Table 2. Effect of the Peptides in the Maximal Rate of Platelet Aggregation Induced by PAF

Comp.	PAF(mol/l)	Rate of Maximal Aggregation (%) Without peptide	With peptide
1	5×10^{-8}	34.86±3.81	15.19±3.26***
2	5×10^{-8}	29.76±4.94	29.58±5.41
3	5×10^{-8}	29.76±4.94	25.39±2.97
4	5×10^{-8}	26.82±1.61	26.82±1.98
5	5×10^{-8}	21.87±3.19	17.99±1.47
6	5×10^{-8}	22.84±1.91	17.79±1.28**
7	5×10^{-8}	22.51±3.62	22.81±1.95

Compare to no peptide ** $P<0.01$; ***$P<.001$, n=4.

References

1. Brandt-Rauf, P.W., J. Protein Chemistry, 8 (1989) 149.
2. Sun, X., J. Cell Biol., 118 (1992) 693.
3. Kawasaki, K., Peptides, ESCOM, 1993 p.120.

Studies on the hybrid peptides of fibrinogen fragments

Ming Zhao, Shi-Qi Peng and Yin-Ye Wang
School of Pharmaceutical Sciences, Beijing Medical University, Beijing 100083, China

Introduction

P6A (ARPAK) is known to increase microvascular permeability and coronary blood flow, improve dynamics in dogs with coronary thrombosis, and induce dilation of bovine mesenteric arteries. QP6A (QRPAK) exhibited a more potent effect than P6A. RGD containing peptides are able to bind to GPIIb/IIIa and inhibit platelet aggregation. The present paper reports the synthesis and linkage peptides and the antithrombotic effects of the results of hybrids of P6A, QP6A and RGDS, RGDV, RGDF.

Results and Discussion

Under the conditions of solution method ARPAKRGDS, ARPAKRGDV, ARPAK-RGDF, QRPAKRGDS, QRPAKRGDV, and QRPAKRGDF were obtained in excellent yields (Fig. 1 and Table 1). The results indicated that the chemical combination of P6A and RGD enhanced the antithrombus effect (Table 2 and 3).

A(Q)+{R+[P+(A+K)]}→ AR(Tos)PAK(Z) or QR(Tos)PAK(Z)→ ARPAK or QRPAK
+ } → 1 or 2 or 3 or 4 or 5 or 6→ 7 or 8 or 9 or 10 or 11 or 12
R+{G+[D+S(V,F)]} → RGDS or RGDV or RGDF

AlaArg(Tos)ProAlaLys(Z)Arg(Tos)GlyAsp(OcHex)Ser(Bzl)OBzl	(1)
GlnArg(Tos)ProAlaLys(Z)Arg(Tos)GlyAsp(OcHex)Ser(Bzl)OBzl	(2)
AlaArg(Tos)ProAlaLys(Z)Arg(Tos)GlyAsp(OcHex)ValOBzl	(3)
GlnArg(Tos)ProAlaLys(Z)Arg(Tos)GlyAsp(OcHex)ValOBzl	(4)
AlaArg(Tos)ProAlaLys(Z)Arg(Tos)GlyAsp(OcHex)PheOBzl	(5)
GlnArg(Tos)ProAlaLys(Z)Arg(Tos)GlyAsp(OcHex)PheOBzl	(6)
AlaArgProAlaLysArgGlyAspSer	(7)
GlnArgProAlaLysArgGlyAspSer	(8)
AlaArgProAlaLysArgGlyAspVal	(9)
GlnArgProAlaLysArgGlyAspVal	(10)
AlaArgProAlaLysArgGlyAspPhe	(11)
GlnArgProAlaLysArgGlyAspPhe	(12)

Fig. 1. The Synthetic Route.

Table 1. The Yield and FAB-MS of Compounds 1-12

Comp.	1	2	3	4	5	6
Yield(%)	95	93	96	94	95	93
MS[M+Na]$^+$	1783	1840	1705	1762	1753	1800
Comp.	7	8	9	10	11	12
Yield(%)	75	76	76	77	75	76
MS[M+Na]$^+$	979	1036	991	1049	1039	1096

Table 2. The Amino Acid Analysis of Compounds 7-12

	Amino acid residue									
Comp.	Ala	Pro	Lys	Glu	Arg	Gly	Asp	Ser	Val	Phe
7	1.98	0.97	0.97	--	1.96	0.98	1.01	0.96	--	--
8	1.97	0.98	0.97	--	1.98	1.01	0.97	--	1.01	--
9	1.99	1.00	0.98	--	1.97	1.00	0.98	--	--	1.01
10	--	1.01	0.97	0.99	1.98	0.99	0.99	0.97	--	--
11	--	0.97	0.99	0.97	1.98	0.98	0.98	--	0.99	--
12	--	0.98	0.97	0.98	1.98	0.99	0.97	--	--	0.98

Table 3. Effect of Hybrid Peptides on Thrombosis

Comp.	Dose	The weight of thrombus(mg)	Comp.	Dose	The weight of thrombus(mg)
NS	-	0.67±0.12	7	2.5	0.45±0.28*
ARPAK	5.0	0.34±0.22**	8	2.5	0.48±0.10*#
QRPAK	2.5	0.44±0.08**	9	2.5	0.45±0.15*#
RGDS	5.0	0.61±0.23	10	2.5	0.41±0.11**#
RGDV	5.0	0.59±0.16	11	2.5	0.71±0.17
RGDF	2.5	0.41±0.26*	12	2.5	0.44±0.37

Dose: µmol/kg; n=6; compare to NS, *$P<0.05$; ** $P<0.01$; compare to RGDS, RGDV, # $P<0.05$.

References

1. Nichols,A.J., Ruffolo,R. R. and Huffman, Jr. W. F., Tips, 13 (1992) 43.
2. Saiki,I. J., Murata, J. And Matsuno,K., Jpn. J.Cancer Res., 81 (1990) 660.

The combination and function of YIGSR and YIGSK

Li-Ping Tu[a], Karl Specht[a], Li Zhao[a], Zi-Lin Chen[a], Chao Wang[b], Dilinur Sabitouf[b], Ming Zhao[b] and Shi-Qi Peng[b]

[a] *Jiangxi-OAI Gemeinsames Institut, Nanchang 330047, China*

[b] *School of Pharmaceutical Sciences, Beijing Medical University, Beijing 100083, China*

Introduction

In the studies on metastasis related to laminin receptor YIGSR and YIGSK were identified as playing a critical role in binding of protein laminin to the coresponding receptor. It was found that the terminal amino acid in the laminin pentapeptide produced significant structural changes. In the present paper YIGSR (1) and YIGSK (2) were combined to yield YIGSRYIGSK(3) and YIGSKYIGSR(4) to compare the bioactivity change induced by terminal and central amino acid change.

Results and Discussion

Peptides 1-4 were synthesized with protected L-amino acids by use of solution method (Fig. 1). Their antiaggregation effects show no significant differences, suggesting that as platelet aggregation inhibitors they were not structural dependent (Table 1 and Table 2).

Table 1. Effect of the Peptides in the Maximal Rate of Platelet Aggregation Induced by ADP (n=4).

Comp.	ADP(mol/l)	Rate of Maximal Aggregation (%) Without peptide	With peptide
1	1×10^{-5}	28.24±3.14	27.01±2.21
2	5×10^{-6}	28.29±2.97	25.48±1.90*
3	1×10^{-5}	40.52±5.44	33.93±1.10
4	1×10^{-5}	40.52±5.44	37.69±3.27

Y+{I+[G+(S+R)]}—— Y(Bzl)IGS(Bzl)R(Tos)—— YIGSR(1)
+ }YIGSRYIGSK(3),YIGSKYIGSR(4)
Y+{I+[G+(S+K)]}—— Y(Bzl)IGS(Bzl)K(ClZ)—— YIGSK(2)

Fig. 1. The Synthetic Route

Table 2. Effect of the Peptides in the Maximal Rate of Platelet Aggregation Induced by PAF (n=4).

Comp.	PAF(mol/l)	Rate of Maximal Aggregation (%) Without peptide	With peptide
1	5×10^{-8}	26.82±1.61	26.82±1.98
2	5×10^{-8}	20.22±3.36	19.65±4.06
3	5×10^{-8}	35.66±2.92	33.21±4.50
4	5×10^{-8}	35.66±2.92	34.37±3.31

References

1. Bramadt-Rauf, P.W., J. Protein Chemistry, 8 (1989) 149.
2. Sun, X., J. Cell Biol., 118 (1992) 693.
3. Kawasaki, K., Peptides ESCOM, 1993, p. 120.
4. Murthy ,M.S., Clin. Exp. Metastasis, 10 (1992) 39.
5. Saiki, I., Jpn. J. Cancer Res., 81 (1990) 660.

The relaxation effect of the linkers of APLRV and RGD

Ming Zhao, Shi-Qi Peng and Dilinur Sabitouf
School of Pharmaceutical Sciences, Beijing Medical University, Beijing 100083, China

Introduction

Studies on Fgn binding to GPIIb/IIIa have identified two distinctive amino acid sequences within the Fgn molecule that mediate its attachment to GPIIb/IIIa. These sequences are Arg-Gly-Asp-Ser (RGDS) and Arg-Gly-Asp-Phe (RGDF) which correspond to 572-575 and 95-98 in the Fgn A α chain respectively. They are also present in FN and vWF and appear to mediate the binding of each to GPIIb/IIIa.

The binding function of GPIIb/IIIa can be blocked by synthetic (small molecular) peptides containing RGD and APLRV, resulting in the prevention of platelet-dependent thrombosis. It was found that the synthetic peptide was also vasodilator. In an attempt to demonstrate their vasodilation effect RGDS, RGDF, APLRV, APLRVRGDS and APLRVRGDF were investigated.

Results and Discussion

RGDS, RGDF, APLRV, APLRVRGDS and APLRVRGDF were synthesized with protected L-amino acids using solution method (Fig. 1).

Relaxation effects were evaluated on aortic strips treated with noradrenaline(NE) (Fig.2-6). The results indicated that all of the peptidesexcept RGDF exhibited vasodilation effects dose-dependently. Among them APLRVRGDS was the best vasodilator. The data suggested that coupling APLRV with RGDS or RGDF may improve their activity and APLRV may be responsible for the increase of the vasodilation effect of APLRVRGDS.

Ala+{Pro+[Leu+(Arg+Val)]}→ APLRV
+ }→ APLRVRGDS or APLRVRGDF
Arg+{Gly+[Asp+Ser(Phe)]}→ RGDS(F)

Fig. 1. The synthetic Route.

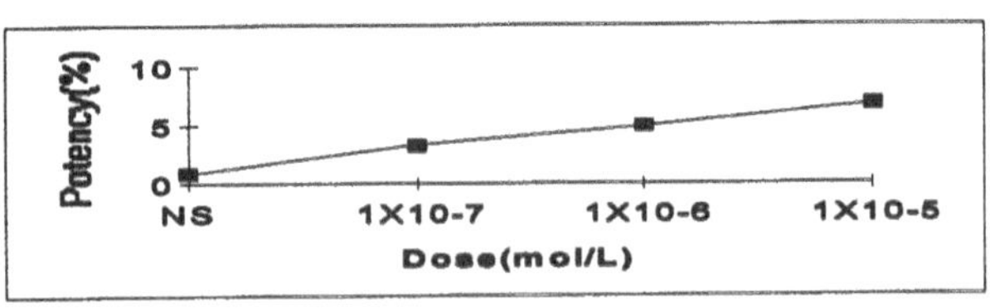

Fig. 2. The vasodilation effect of APLRV

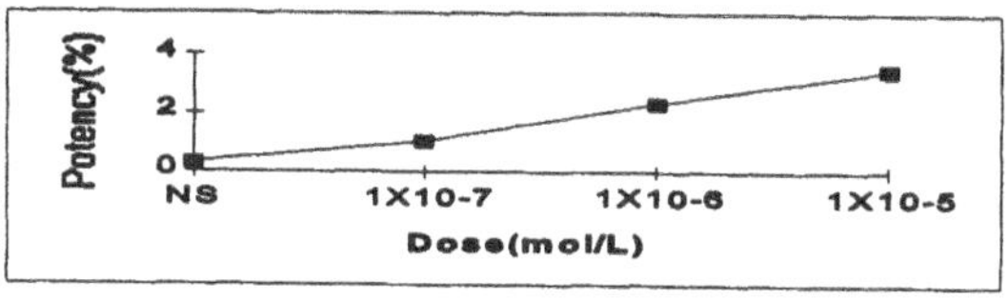

Fig. 3. The vasodilation effect of APLRVRGDS

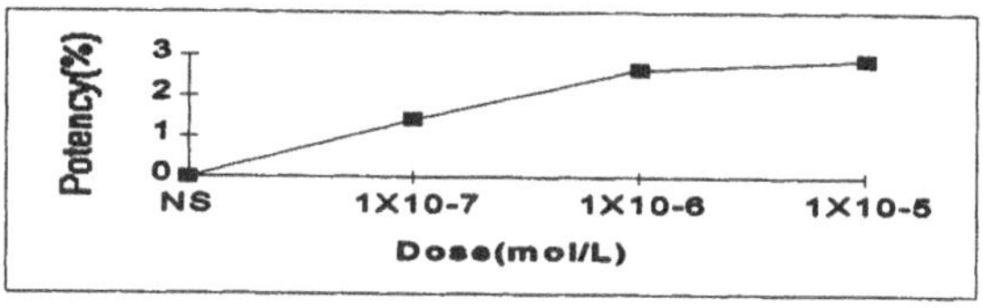

Fig. 4. The vasodilation effect of RGDS

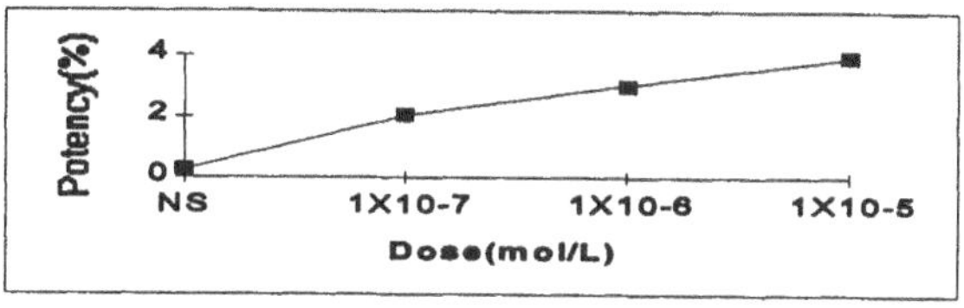

Fig. 5. The vasodilation effect of APLRVRGDF

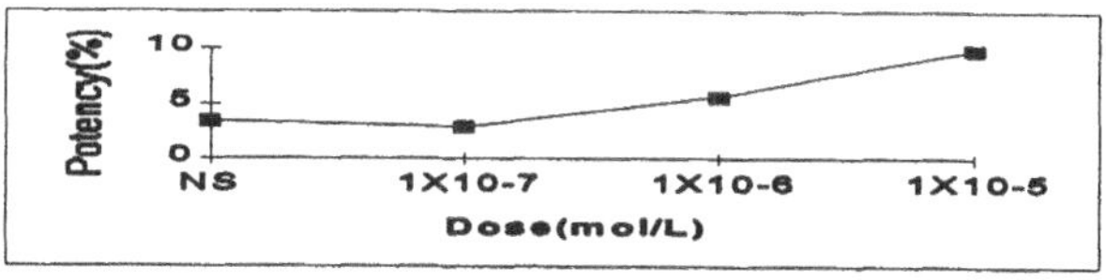

Fig. 6. The vasodilation effect of RGDF

References

1. Gartner ,T.K. and Taylor ,D.B., Supplement Circulation, 78 (1988) 2480.
2. Nichols, A.J., Ruffolo, R.R. and Huffman Jr. W.F., Tips, 13 (1992) 413.

Studies on the linkers of steroids and peptides

Chao Wang, Shi-Qi Peng and Xue-Cai Qiu

School of Pharmaceutical Sciences, Beijing Medical University, Beijing 100083, China

Introduction

In studies of the interaction between steroid and peptide hormones hydrocortisone was linked with UTP-A (His-Gly-Lys), UTP-B (Glu-Asp-Gly) and UTP-C (His-Gly-Glu), respectively. The linkers, hydrocortisone-UTP-A(1), hydrocortisone-UTP-B (2), hydrocortisone-UTP-C(3) and UTP-A-hydrocortisone(4) were obtained.

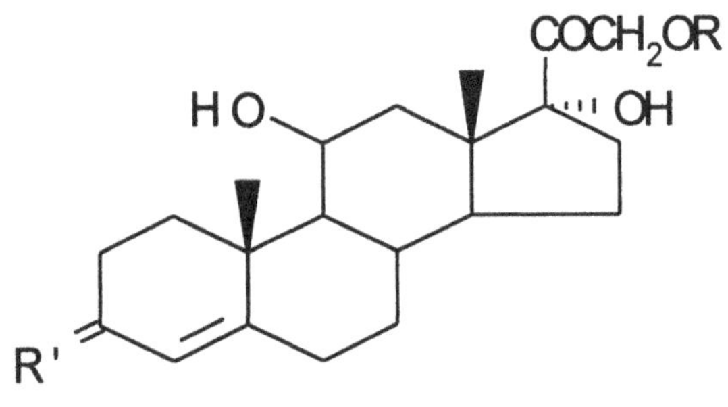

1. R'=O, R=$COCH_2CH_2CO$-His-Gly-Lys-OH

2. R'=O, R=$COCH_2CH_2CO$-Glu-Asp-Gly-OH

3. R'=O, R=$COCH_2CH_2CO$-His-Gly-Glu-OH

4. R'=H-His-Gly-Lys-NH-N=, R=H

Results and Discussion

Immunodepression effects were evaluated by heterotopic transplant of cardiac tissue of fetal mouse, phagocutosis of peritoneal mouse macrophages, and mitogen- induced proliferation of mouse spleen cells (Tables 1-3). The results indicated that the linkers always exhibited higher potency than steroid or peptide,suggesting that hormone steroid and peptide hormones can enhance their bioactivities.

Table 1. Prolongation of heterotopic transplanted cardiac tissue survival induced by compounds 1, 2, and 7

Comp.	Mean graft survival (days, X±E)		
dose(mmol/20g/d)	0.15	0.75	1.50
control	8.87±0.03	8.87±0.03	8.87±0.03
Hydrocortisone	9.86±0.57	10.86±0.67**	11.67±0.59***
UTP-A	9.25±0.41	10.02±0.51*	10.30±0.56
Comp. 3	10.87±0.37**#	11.00±0.49**	11.78±0.64***

n=9; Compare to control*P<0.05, **P<0.01, ***P<0.001; Compare to 2 and 4 #P<0.05.

Table 2. The Inhibitory effects of compounds 1, 2, 4, 5, 7 and 8 on phagocytosis of mouse penitoneal macrophages(n=12)

	Optical density (X ± SE) at different concentration (mol/l)				
Comp.	0.001	0.010	0.100	10.00	1.000
Hydrocordisone	0.66±0.03*	0.56±0.04***	0.53±0.05***	0.46±0.03***	0.50±0.06***
UTP-A	0.74±0.02	0.65±0.05*	0.63±0.03**	0.59±0.03***	0.64±0.04*
UTP-C	0.81± 0.03	0.66±0.02*	0.62±0.05***	0.54±0.03***	0.55±0.04***
1	0.68±0.04	0.52±0.02***	0.45±0.03***	0.41±0.03***	0.40±0.05***
3	0.55±0.03***	0.52±0.04***	0.41±0.02***	0.38±0.04***	0.45±0.06***
4	0.65±0.03*	0.53±0.04***	0.44±0.03***	0.41±0.04***	0.44±0.05***

Control(0 μ mol/l): 0.78±0.04; Compare to control * P<0.05, ** P<0.01, *** P<0.001.

Table 3. The Influence of compouds 1, 2, 4, 5, 7 and 8 on ConA induced proliferation of spleen lymphocytes of mouse(n=6)

	inhibitory rate (X ± SE) at defferent concentration (mol/l)				
Comp.	10-8	10-7	10-6	10-5	10-4
Hydrocortisone	6.3±5.1	36.0±5.2**	56.4±4.2**	80.6±4.6**	75.9±10.5**
UTP-A	5.0±5.3	22.4±5.3*	48.9±16.3**	55.7±4.2**	65.2±7.2**
UTP-C	8.8±4.2	34.5±3.5**	37.4±18.6**	62.4±4.1**	64.3±8.2**
1	29.3±5.9**$	31.4±5.1**	75.6±5.0**	79.8±11.6**	89.2±16.1**
3	37.8±3.6**##	51.2±3.1**	77.9±5.7**	84.7±5.5**	77.0±5.7**
4	30.7±4.6**$$	44.0±5.4**	71.8±7.1**	78.3±4.4**	79.2±11.1**
Hydrocortisone+UTP-A	5.0±6.0	35.0±4.7**	72.4±7.2**	70.5±4.5**	78.4±8.3**
Hydrocortisone+UTP-C	7.5±5.5	35.0±6.3**	66.7±10.4**	79.3±7.7**	76.3±4.5**

The inhibitory rate of control was 0.00; compare to control *P<0.05,** P<0.01; compare to Hydrocortisone+UTP-A, $P<0.05; $$P<0.01; compare to ## Hydrocortisone+UTP-C, P<0.01.

References

1. Ingle, D. J., Proceedings of the Society of Endocrinology, J. Endocrinol., 8 (1952) 29.
2. Zhu,Y., Acta Pharm. Sinica, 28 (1993) 166.
3. Chen, X.-H. Bulletin, 5 (1989) 78.
4. Biko, T., Biochem. Biophys Res. Commun., 86 (1979) 86.

A novel series of ET_A selective antagonists

Jia-Ning Xiang, Steven T. Atkinson, Dimitri Gaitanopoulos, Deborah Bryan, Ponnal Nambi, Eliot H. Ohlstein, Joseph Weinstock and John D. Elliott
SmithKline Beecham Pharmaceuticals, King of Prussia, PA 19406, U.S.A.

Introduction

The endothelins (ETs) are a family of three isopeptides (ET-1, ET-2 and ET-3) with potent vasoconstrictor properties [1]. All three of these peptides are encoded in the human genome and each is composed of 21 amino acids, with disulfide linkages between cysteine residues at positions 1,15 and 3,11. The ETs elicit their effects through binding to receptors of the G-protein coupled seven-transmembrane spanning superfamily. Two receptor subtypes have thus far been fully characterized from human tissues through molecular cloning and expression. Recently, compelling data has become available using receptor specific antagonists to support a role for the ETs in disease processes. Despite extensive animal model studies, however, controversy still surrounds the optimal antagonist selectivity considered necessary for the treatment of human disorders [2] and this may also be dependent upon the disease targeted.

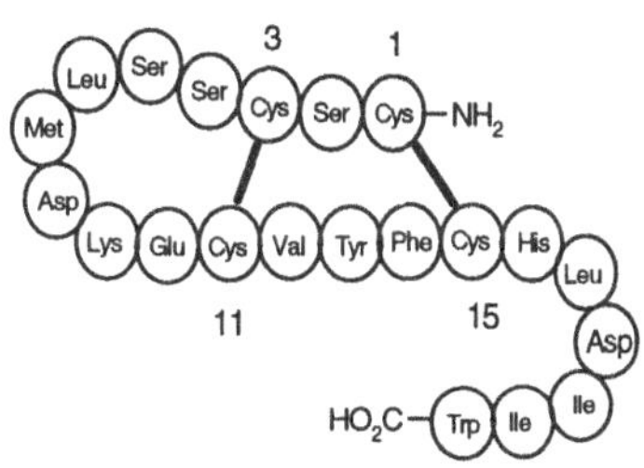

Endothelin-1

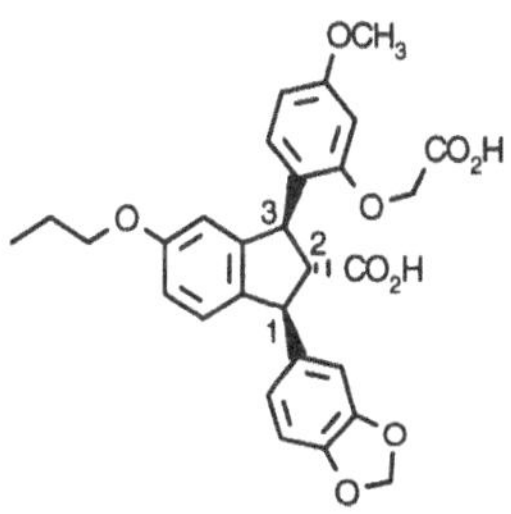

1 SB 209670

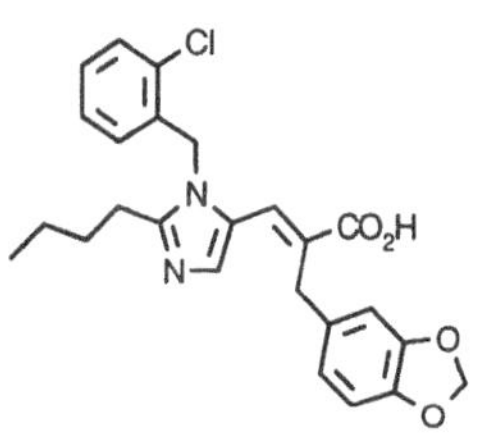

2 SK&F 107328

Results and Discussion

Recently, we reported the design and synthesis of 1,3-diarylindane-2-carboxylic acids as potent endothelin antagonists, exemplified by SB 209670 (**1**) [3]. SB 209670 (**1**) is a highly potent antagonist which inhibits [^{125}I]ET-1 binding to cloned human ET_A and ET_B receptors with K_i values of 0.43 and 14.7 nM, respectively. Structure-activity relationship studies in the indane series highlight the importance of the 2-carboxyl and the benzodioxole moieties for high affinity binding. Concurrent with our work in the indane series, a chemical database searching was conducted for acidic compounds containing an appropriately connected benzodioxole moiety (i.e. a carboxylic acid linked by two atoms to the 5-position of a benzodioxole group) and *via* this approach an additional lead structure SK&F 107328 was identified. SK&F 107328 (**2**), originally prepared for our angiotensin II receptor antagonist program, is a dual antagonist of ET and AT-1 receptors [K_i's: ET_A= 400 nM, ET_B= 3400 nM, AT-1= 180 nM). Molecular modeling overlays in which the 2-carboxylic acid and benzodioxole groups of SB 209670 are superimposed upon the acrylic acid and benzodioxole moieties of SK&F 107328 fail to show a correspondence of the 3-phenyl of SB 209670 and the N-chlorobenzyl moiety of SK&F 107328. However, it is apparent that a good overlay can be achieved with an analog of SK&F 107328 in which a phenyl substituent corresponding to the indane 3-phenyl is placed at the 3-position of the acrylic acid. At the same time, known SAR in the angiotensin receptor antagonist area suggested that removal of the N-chlorobenzyl group from SK&F 107328 would diminish affinity to the AT-1 receptor. Thus, SB 209834 (**3**) was prepared which embodies these modifications.This compound is indeed a more potent and selective antagonist of the endothelin receptors (K_i's: ET_A= 2 nM, ET_B= 500 nM, AT-1>10 μM) [4].

While SB 209834 does not contain any chiral centers the geospecific synthesis of the tetra-substituted olefin presents a significant synthetic challenge. Our attention therefore turned toward modifications of this molecule which might further simplify the structure.

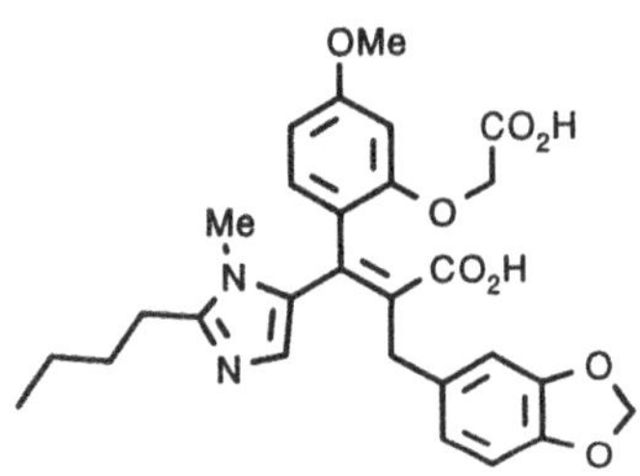

3 SB 209834

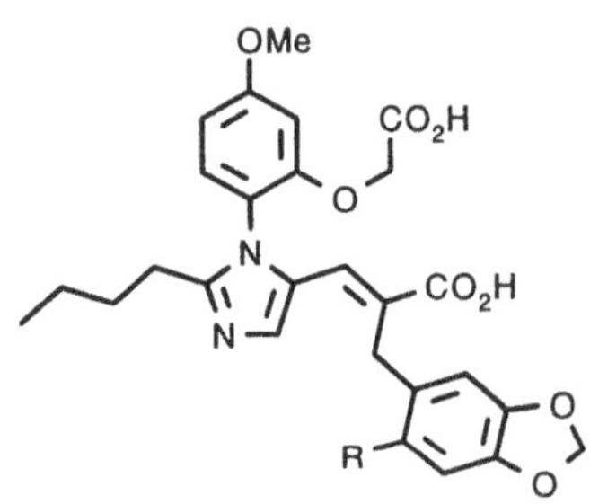

4 SB 212954 R=H
5 SB 215355 R=OMe

Somewhat surprisingly, it was discovered that the phenyl group at position 3 in SB 209834 could be translocated to N-1 of the imidazole to provide SB 212954 (**4**), an analog of almost identical affinity to ET_A and slightly greater selectivity versus ET_B [K_i's: ET_A= 2 nM, ET_B= 625 nM, AT-1> 10 μM). While the beneficial effect upon binding of the oxyacetic moiety on the N-phenyl ring in this series is not as pronounced as in the indane analogs [3], it is superior by at least an order of magnitude in affinity over other substituents studied e.g. OMe, OH, Cl. In common with the indane series the benzodioxole portion of SB 212954 is critical to receptor binding (Table 1). In the N-phenylimidazole series further improvement of affinity is achieved when a 2-methoxy substituent is added to the benzodioxole moiety generating the potent ET_A selective antagonist SB 215355 (**5**) [K_i's ET_A= 0.7 nM, ET_B= 1 μM]. In functional assays, SB 215355 blocks the contractile response to ET-1 at ET_A (rat aorta) with K_b value of 3.3 nM and yields a Schild analysis consistent with competitive antagonism.

The synthesis of SB 215355 is shown in Scheme 1. Thus, commercially available 2,4-

dichloronitrobenzene is converted to 2,4-dimethoxynitrobenzene by treatment with sodium methoxide in methanol in 91% yield. Selective 2-demethylation is achieved by reaction

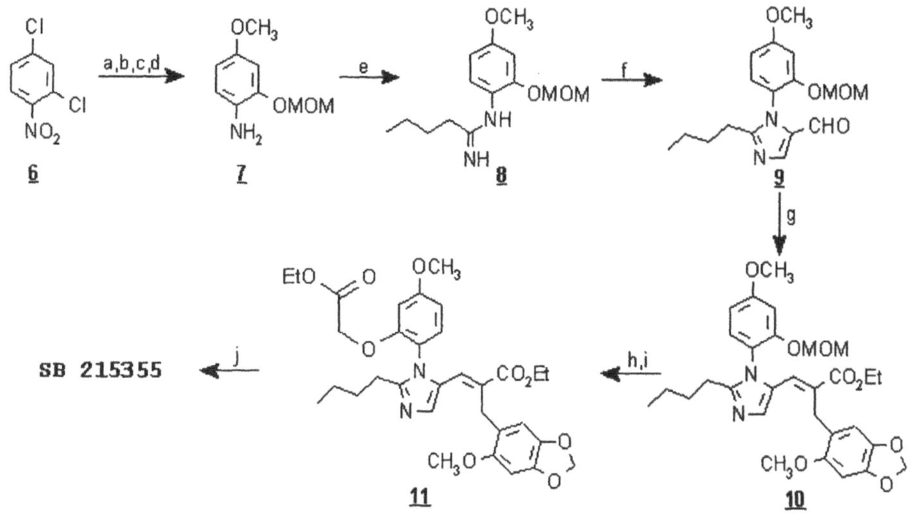

a) NaOMe/MeOH, reflux (91%); b) $MeSO_3H$/*dl*-Methionine, r.t. (84%); c) NaH/DMF/MOMBr (88%); d) H_2, 10%Pd/C, EtOH (95%); e) 1-Methoxy-1-iminopentane/CH_2Cl_2 (85%); f) 2-Bromomalonidialdehyde TEA, AcOH, i-PrOH, reflux (63%); g) 2-Carboethoxy-3-(2-methoxy-4,5-methylenedioxyphenyl)propionic acid/piperidine/AcOH/benzene (63%); h) HCl/MeOH, reflux (93%); i) NaH/DMF/$BrCH_2CO_2Et$; j) NaOH/H_2O/MeOH then HCl/H_2O (80%, two steps).

with DL-methionine in the presence of methanesulfonic acid. Reprotection as the MOM ether gives, after catalytic hydrogenation of the nitro moiety, aniline **7** in 64% yield.

Amidine formation with 1-methoxy-1-iminopentane and subsequent conversion to the desired imidazole nucleus through reaction with 2-bromomalondialdehyde gives **9** in 54% overall yield.

Knoevenagel condensation of **9** with 2-carboethoxy-3-(2-methoxy-4,5-methylenedioxyphenyl) propionic acid followed by removal of the MOM ether protection under acid conditions, and introduction of the oxyacetic ester side chain *via* alkylation with ethyl bromoacetate affords diester **11**. Finally, deprotection of the diester **11** by saponification followed by acidification yields SB 215355 (47% yield from **9**).

In conclusion, we have designed a novel ET_A selective antagonist SB 215355 of high potency, which is without measurable affinity to the AT-1 receptor. This compound should be a valuable tool with which to probe the involvement of the ET_A receptor subtype in disease processes.

References

1. Yanagisawa, M., Nature 332 (1988) 411.

2. Ruffolo Jr, R.R. (Ed.), Endothelin Receptors: From the Gene to the Human, CRC Press Inc., Boca Raton, Fl, U.S.A., 1995.
3. Elliott. J.D., J. Med. Chem., 37 (1994) 1553.
4. Elliott, J.D.,In Kaumaya, P.T.P. and Hodges, R.S. (Eds.), Peptides: Chemistry, Structure and Biology (Proceedings of the 14th American Peptide Symposium), Mayflower Scientific Ltd., 1996, p. 673.

Syntheses of insulins: From *in vivo* to *in vitro* to *in vivo*

You-Shang Zhang
Shanghai Institute of Biochemistry, Chinese Academy of Sciences, Shanghai 200031, China

Insulin was first isolated from the canine pancreas by Banting and Best in 1921. On the occasion of the 75th anniversary of this great discovery in biology and medicine, it is appropriate to review briefly studies on insulin syntheses in which my colleagues and I have been involved. As a protein hormone, insulin is synthesized *in vivo* in the islet cells of animals. Insulin dependent diabetics relied solely on insulin isolated from the pancreas of animals such as cattle and pig until recombinant human insulin became available in recent years. Porcine insulin, which is mainly produced in China, turns out to be the nearest relative of human insulin and has been used extensively in the clinic. Over many years, insulin from animal pancreas continues to be isolated through acid ethanol extraction. For removing ethanol, different methods are available, such as vacuum distillation, ion exchange adsorption or calcium phosphate gel adsorption, which was used in our laboratory many years ago for the large scale production of porcine insulin as well as the preparation of insulins from different species [1]. The calcium phosphate gel was formed in situ at pH 6 from phosphoric acid present in the extract and calcium chloride added later. The adsorbed proteins containing insulin were eluted in aqueous acid solution. Using this method, rhombohedral chicken insulin crystals and dodecahedral snake insulin crystals were obtained. Chicken insulin shows higher *in vitro* activity than insulins from other species and crystals of chicken and porsine insulins are isomorphous. It is worthwhile to solve the three-dimensional structure of chicken insulin and determine the structural basis of its high activity.

With the progress of peptide synthesis, chemical synthesis *in vitro* became a challenge in the Fifties. In the Sixties, bovine or ovine insulin was synthesized in laboratories of three countries including China. The resynthesis of native insulin in crystalline form was of great help [2]. It simplified the synthesis of the whole molecule to the syntheses of its two separate chains. It also told us that with the unique amino acid sequence and correct disulfide bond pairing of A and B chains, the insulin molecule would fold in the right native way. Semisynthesis, especially with the aid of enzymes, has advantages for studying structure and function relationships of insulin. Semisynthesis can be considered as a combination of *in vivo* and *in vitro* syntheses using a fragment of insulin produced *in vivo* coupled with a fragment synthesized *in vitro*. Using tryptic semisynthesis, many analogues with shortened or modified C-terminal regions of the B chain were prepared. Among them is deshexapeptide insulin (DHI), i.e. insulin with B25 to B30 removed, which is an active fragment of insulin worth mentioning [3, 4]. Its *in vivo* biological activity (40%) is much higher than its binding activity (2%), indicating that the active site and binding site are relatively independent. As Peter Schiller pointed out, analogues with high biological activity and low binding activity are super agonists. Recently, the three-dimensional structure of DHI was solved [5]. It was shown that DHI retained the conformation that was present in DPI or insulin. This result further strengthens the idea that B25 Phe or other residues as well as the lost part may be involved in receptor binding but with little effect on biological activity.

With the progress of gene cloning and site-directed mutagenesis, recombinant human insulin and its mutants were synthesized *in vivo* in micro-organisms, first in *E. coli* by Eli Lilly Co. and later in yeast by Novo Nordisk Co., as reported by Markussen and co-workers [6]. In our laboratory, yeast cells transformed by a plasmid with ADH1 or PGK promoter, alpha-mating factor leader sequence and synthetic gene of insulin precursor have been used for the preparation of human insulin [7] and its mutants with fast acting or long acting activity. For example, we have prepared GluA17Lys human insulin which has reduced (about 10%) biological activity and receptor binding activity but can be crystallized under conditions similar to those used for insulin, indicating that A17 Glu may be involved in the binding site or active site [8]. According to the cross-linking binding model proposed by De Meyts [9], residue A17 in the hexamer-forming surface is involved in the binding of insulin to its receptor. We have also prepared human insulin mutants with B12 Val replaced by Thr and Leu [10]. Surprisingly, B12 Thr mutant has full *in vivo* biological activity while B12 Leu mutant has much lower activity, suggesting that hydrophobicity at B12 may not be essential for insulin activity. Compared to enzymatic semisynthesis, site-directed mutagenesis has the advantaget of changing residues of insulin at any desired position . We believe that deeper insight into the structural biology of insulin will be gained through this powerful approach.

References

1. Cao, Q.P., Li, T.F., Peng, X.H. and Zhang, Y.S., Sci. Sin., 23 (1980) 1309.
2. Du, Y.C., Zhang, Y.S., Lu, Z.X. and Tsou, C.L., Sci. Sin., 10 (1961) 84.
3. Cao, Q.P., Cui, D.F. and Zhang, Y.S., Nature (London), 292 (1981) 774.
4. Zhang, Y.S., Biochem. Transactions, 11 (1983) 417.
5. Chang, W.R., Jiang, T., Ren, Z., Wan, Z.L., Xu, Y.B., Liang, D.C., Zhu, S.Q. and Zhang, Y.S., Science in China (Series B), 38 (1995) 1094.
6. Markussen, J., Diers, I., Engesgaard, A., Hansen, M.T., Hougaard, P., Langkjaer, L., Norris, K., Ribel, U., Sorensen, A.R., Sorensen, E. and Voigt, H.O., Protein Engineering, 1 (1987) 215.
7. Z hang, Y.S., Hu, H.M., Cai, R.R., Feng, Y.M., Zhu, S.Q., He, Q.B., Tang, Y.H., Xu, M.H., Xu, Y.G., Liu, B. and Liang, Z.H., Science in China (Series C), 39 (1996) 225.
8. Zhang, B.Y., Zhang, Y.S., Cai, R.R., Feng, Y.M. and He, Q.B., Biochem. Mol. Biol. International, 363 (1995) 1979.
9. De Meyts, P., Diabetologia, 37 (1994) S135.
10. Wang, Q.Q., Feng, Y.M. and Zhang, Y.S., Biochem. Mol. Biol. International, in press.

Design, synthesis and circular dichroism spectroscopy studies on DNA-binding activities of GCN4(226--252) peptide dimer

He Wang[a], Bin Zhou[b], Yu Luo[a], Zhen-Wei Miao[a], Chang-Qing Chen[b], Xiao-Jie Xu[a] and You-Qi Tang[a]

[a]Department of Chemistry, Peking University, Beijing 100871, China

[b]Shanghai Research Center of Biotechnology, Shanghai 200233, China

Introduction

Basic region-leucine zipper (bZIP) family of DNA-binding protein is characterized as a basic region that contacts DNA and an adjacent region of about 30 residues containing a heptad repeat of leucines. This 'leucine zipper' forms a coiled coil and mediates protein dimerization [1]. Yeast transcriptional activator GCN4 in which the DNA-binding domain locates within the 60 residues of the protein C-terminal is one of the most extensively studied bZIP proteins [2,3]. X-ray structures of two specific GCN4-DNA complexes provide deep insight into the molecular recognition between GCN4 bZIP peptide (226-281) and its sequence-specific DNA target sites [4,5]. In this paper, we have tried to alter the DNA binding specificity of GCN4 bZIP peptide for the purpose of further study on protein-DNA interactions.

Results and Discussion

We selected a peptide model of GCN4(226-252) as our research target, in which the leucine zipper was replaced by a disulfide bond [6].

We have built the computer model of the GCN4-DNA complex from its crystal structure data. Then the peptide MU(Q,Q) was designed, which was predicted to bind the 2 mutated DNA site specifically (Fig.2).

a) oligonucleotides

```
                         -1 0+1
wt   5'-AGAATTCTAGATGACTCATGAATTCCTGCA-3'
       3'-TCTTAAGATCTACTGAGTACTTAAGG-5'      (AP-1 site)
m2   5'-AGAATTCTAGATTACTAATGAATTCCTGCA-3'
       3'-TCTTAAGATCTAATGATTACTTAAGG-5'      ( 2 mutated site)
```

b) peptides

```
              226                           252
WT        Ac-DPAALKRARNTEAARRSRARKLQRLKQGGC-OH
MU(Q,Q)   Ac-DPAALKRARNQEAARRQRARKLQRLKQGGC-OH
```

Fig. 1. The sequences of the peptides and oligonucleotides synthesized. WT and MU(Q,Q) consist of the basic region residues plus the C-terminal linker Gly-Gly-Cys.

Fig. 2. Predicted 3-D structure of MU(Q,Q) dimer-m2 complex.

The 30-residue peptides WT and MU(Q,Q) were synthesized by manual SPPS Fmoc-strategy. Both WT and MU(Q,Q) monomers and their oxidized dimers were purified by RP-HPLC and verified with FAB-MS or MALDI-TOF-MS. Oligonucleotides were synthesized with Applied Biosystems 381A DNA synthesizer, followed by 20% polyacrylamide-urea gel purification.

The DNA-binding activities of WT and MU(Q,Q) dimers were evaluated with CD spectroscopy. The experiments were performed on a Jasco J-720 spectropolarimeter. We measured the helicity of WT and MU(Q,Q) dimer induced by wt or m2 dsDNA, and the thermal stability of four complexes, i.e. WT-wt, WT-m2, MU(Q,Q)-wt and MU(Q,Q)-m2.

The CD spectra showed that the helicity of WT dimer increased 105% upon adding specific wt dsDNA (Fig.3(a)) and increased 80% when induced by m2 (data not shown). WT-wt is the most stable complex among the four peptide-DNA complexes. These experimental results indicated that WT dimer becomes highly helical when bound to native DNA site, but its helicity only partially increases when interacted with 2 mutated DNA site.

The ellipticity of MU(Q,Q) dimer increased 30% in the presence of native DNA site, and its helicity increased only 17% upon addition of m2 dsDNA (Fig.3(b)). Considering the different helix formation of WT and MU(Q,Q) dimers in the absence of DNA (MU(Q,Q) has 26% more helicity than WT), we concluded that the designed MU(Q,Q) dimer failed to bind m2 specifically.

In the crystal structure of GCN4-DNA complex, Ser-242 directly interacts with the 3 thymine methyl, the Gln-242 substitution should affect DNA-binding specificity. Since Gln can recognize base pair A•T by forming two hydrogen bonds with the adenine base, we predicted that Gln-236 and Gln-242 dual substitutions could bind 2 mutated DNA site specifically. But the CD experiments did not show our expected results. We are now involved in the redesigning for the next cycle of design-synthesis-DNA-binding activity assays.

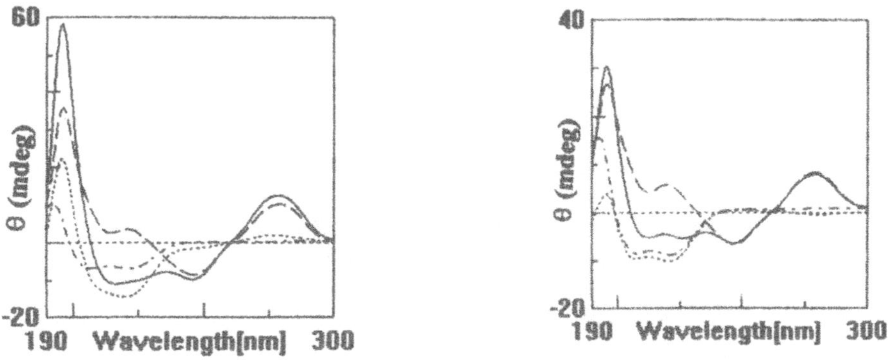

(a) (b)

Fig. 3. CD spectra of peptide dimer in the presence of dsDNA. (a) WT-wt complex; (b) MU(Q,Q)-m2 complex,_____ peptide dimer induced by dsDNA, ------ dsDNA. The difference spectrum was obtained by subtracting dsDNA, ... peptide dimer in the absence of DNA. Conditions: 10mM phosphate buffer(pH7.3), 100mM NaCl, 8.0 °C. Peptide dimer (12 M), DNA (12 M).

References

1. O' Neil, K.T., Hoess, R.H. and DeGrado, W.F., Science, 249 (1990) 774.
2. Kim, J., Tzamarias, D. and Struhl, K., Proc. Natl. Acad. Sci. U.S.A., 90 (1993) 4513.
3. Stanojevic, D. and Verdine, G.L., Nature Structural Biology, 2 (1995) 450.
4. Konig, P. and Richmond, T.J., J.Mol.Biol., 233 (1993) 139.
5. Ellenberger, T.E., Brandl, C.J., Struhl, K. and Harrison, S.C., Cell, 71 (1992) 1223.
6. Talanian, R.V., McKnight, C.J. and Kim, P.S., Science, 249 (1990) 769.

Synthesis, resolution of α-amino-β-[4-(1,2-dihydro-2-oxo-quinolinyl)]propionic acid and its physiological activity

Yun-Hua Ye[a], De-Yi Zhang[a], Jian-Guo Ji[a], Guang-Cheng Jiang[b], Shi-Yi Liu[c], Wen-Yuan Zhang[c] and Wen-Yan Chen[c]

[a] *Department of Chemistry, Peking University, Beijing 100871, China*
[b] *Tianjin Municipal Research Institute for Family Planning, Tianjin 300191, China*
[c] *Shanghai Institute of Physiology, Academia Sinica, Shanghai 200031, China*

Introduction

A nonprotein amino acid, DL-α-amino-β-[4-(1,2-dihydro-2-oxo-quinolinyl)] propionic acid (DL-Aoq) was synthesized in high yield by an improved method and resolved for the first time by chymotrypsin or subtilisin to obtain L-Aoq and D-Aoq. The optical purity of the L- and D-isomers is more than 95% and 99% respectively as detected by a chiral HPLC column. The physiological activities (decrease of respiratory rhythm, decrease of heart rate and delta-enhancing effect) of DL-Aoq, L-Aoq and D-Aoq were evaluated in adult male and female rabbits.

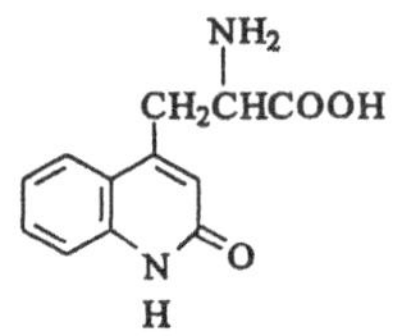

Aoq

Results and Discussion

Synthesis

Instead of NaH/benzene [1], Na/EtOH was used for a condensation reaction between 1,2-dihydro-2-oxo-4-bromomethyl quinoline and diethyl N-acetylamino-malonate in mild condition and short reaction time to obtain diethyl N-acetylamino-α-[4-(1,2-dihydro-2-oxo-quinolinyl)-methyl]-malonate, yield 70%, m.p.: 224~225°C (Lit[1]: yield 48%, m.p.: 225°C). This intermediate compound was hydrolyzed by base, and then acidified, and decarboxylated to obtain DL-Aoq·HCl, yield 84%, m.p.: 252~254°C, DL-Aoq, m.p.: 286~288°C (dec.), (Lit[1]: yield 62%, m.p.: 288°C(dec.)). DL-Aoq was described by ^{1}HNMR, MS, UV, IR and elemental analysis [2] while the Lit[1] only gave its melting point.

Resolution

The key step of resolution was to obtain the monoester, N-Ac-DL-AoqOEt. After many trials, we found that N-Ac-DL-AoqOEt was easy to obtain by diethyl N-acetylamino-α-[4-(1,2-dihydro-2-oxo-quinoline)-methyl]-malonate reaction with 0.5N NaOH at 0°C and stirring for 8h. N-Ac-DL-AoqOEt was then resolved by chymotrypsin (pH=7.5) or subtilisin (pH=8.0) to get N-Ac-L-AoqOH . N-Ac-D-AoqOEt did not hydrolyze in the same condition. After separation and acid hydrolysis, L-AoqOH and D-AoqOH were obtained, respectively. The purity of L- and D-isomers was more than 95% and 99%, respectively, detected by a chiral HPLC column[3]. L-AoqOH·HCl, MS(FAB): 233$(M+H)^+$, m.p.(dec.): 245~248°C, $[\alpha]_D^{20}$: +16.6 (c=1, 0.5N HCl); D-AoqOH·HCl, MS(FAB): 233$(M+H)^+$, m.p.(dec.): 248~252°C, $[\alpha]_D^{20}$: -18.3 (c=1, 0.5N HCl). The physical constants are identical to those compounds that were prepared by asymmetric synthesis and chemical resolution [4]. The yields of our method were higher than Lit[4].

Physiological activities

The physiological activities of both D-Aoq and L-Aoq were evaluated in adult rabbits of either sex. The delta index, sigma index, as well as respiratory rhythm and heart rate were served as physiological variables by using mainly mesodiencephalic intraventricular infusion (i.c.v.), though intravenous administration (i.v.) was also adopted for comparison. Results in rabbits showed that the mesodiencephalic infusion of both D-Aoq and L-Aoq (10 μg /animal, i.c.v.) could exert similar significant central inhibitory effects to that of intravenous administration (100 μg/Kg, i.v.). It is of interest that though the sigma-enhancing effects of both D-Aoq and L-Aoq (10 μg /animal, i.c.v.) were significant ($P<0.001$), the delta-enhancing effects of L-Aoq [175 ± 20 (N=8) vs 114 ± 9 for the control (N=6), $P<0.05$) were less significant when compared with D-Aoq [202 ± 19(N=8) vs 114 ± 9 for the control (N=6), $P<0.001$]. On the other hand, the mesodiencephalic intraventricular infusion of both D-Aoq and L-Aoq (10 μg/animal) could also decrease respiratory and heart rate in rabbits. Nevertheless, although the effects of decreasing respiratory rates per min of both D-Aoq and L-Aoq (10 μg/animal, i.c.v.) were significant ($P<0.001$), the effects of decreasing heart rate per min were significant only for L-Aoq [93 ± (N=8) vs 96±1 for the control (N=6), $P<0.001$] and not D-Aoq ($P>0.05$).

Acknowledgments

This work was supported by Doctoral Program Foundation of Institute of Higher Education and National Natural Science Foundation of China.

References

1. Shak, D.O. and Trived, K.N., Cur. Sci., 41 (1972) 876.
2. Ye, Y.H., Ji, J.G., Qiu, X.S., Liu, S.Y., Zhang, W.Y., Wang, Z.S. and Dai, X.J., Acta Scientiarum Naturalium Universitatis Pekinensis, 26 (1990) 207.
3. Ye, Y.H., Zhang, D.Y., Ji, J.G., Liu, S.Y., Zhang, W.Y. and Dai, X.J., Reported at the 3r d Sino-French Symposium on the Chemistry of Natural Products (October 5~7, 1993, Shanghai, China) p. 55.
4. Otsubo, K., Morita, S., Uchida, M., Yamasaki, K., Kanbe, T. and Shimizu, T., Chem. Pharm. Bull., 39 (1991) 9061.

Session VI
Chemistry-Biology-Interface

Chairs: Jurgen Engel
ASTA Medica AG
Frankfurt, Germany

Saburo Aimoto
Osaka University
Osaka, Japan

You-Shang Zhang
Shanghai Institute of Biochemistry
Shanghai, China

and

Yu-Cang Du
Shanghai Institute of Biochemistry
Shanghai, China

Conformational changes of bovine pancreatic polypeptide (BPP) as it interacts with lipids

Song-Qing Nie[a], Li Yang[a], Ke-Chun Lin[a], Yue Zhou[b], Sen-Fang Sui[b] and Hou-Li Jiang[c]

[a]*Department of Biophysics, Beijing Medical University, Beijing 100083, China*

[b]*Department of Biological Science and Technology, Tsinghua University, Beijing 100084, China*

[c]*National Center of Biomedical Analysis, Academy of Military Medical Science, Beijing 100850, China*

Introduction

BPP is a linear polypeptide composed of 36 amino acid residues secreted by the pancreas [1]. Experiments *in vivo* demonstrated its function of treating and preventing acute pancreatitis [2]. Our research showed the effects of BPP on stabilizing the lipid bilayer structure and inhibiting the membrane fusion of liposomes [3]. After interaction with lipids conforma-tional changes of BPP were studied by monolayer, fluorescence and FT-IR spectroscopy techniques.

Results and Discussion

The results of monolayer experiments showed that the surface membrane pressure of lipids increased with added BPP. π-$\Delta\pi$ curves for various lipids showed that the critical membrane pressure (π_c) for CL, DOPA, DOPC, DOPE and DPPC was 38mN/m, 32mN/m, 28mN/m, 26mN/m and 23mN/m respectively (Fig.1), indicating the interaction of BPP with lipid and its penetration into the membrane was stronger for negatively charged lipids than neutral lipids.

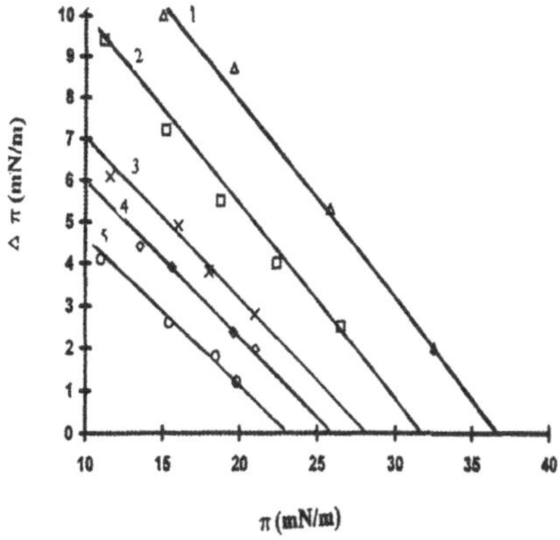

Fig. 1. $\Delta\pi$-π curve of interaction of BPP with CL(1),DOPA(2),DOPC(3), DOPE(4), DPPC(5). conc. of BPP 2μg/ml,pH 7.5,T. 22 °C.

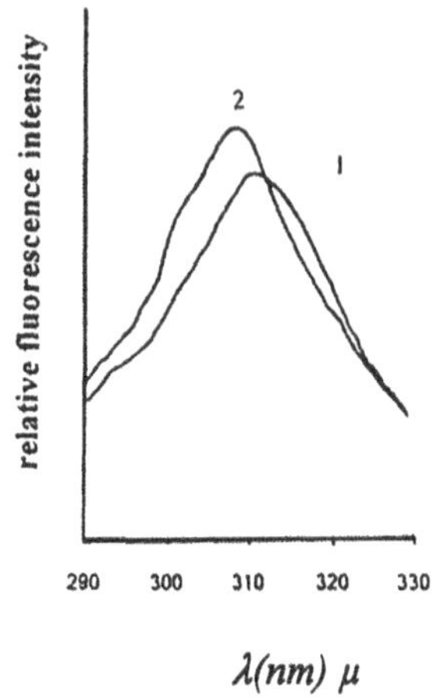

Fig. 2. 1. fluorescence spectrum of fluorescence PP conc. 2.38 μ mol/l. 2. adding DMPC liposomes to BPP. BPP/lipid 1:100 λ_{ex}=280 nm.

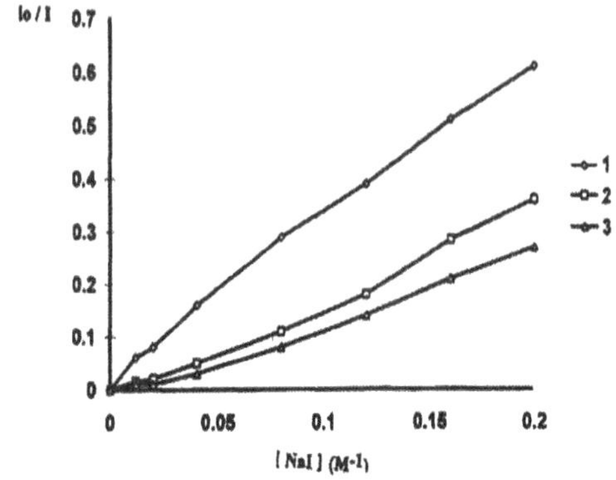

Fig. 3. Stern-Volmer curve of fluorescence quenching of Tyr on BPP by NaI. λ_{ex}=280 nm. λ_{em}=308nm 1. BPP solution. Ksv 3.0 mol/l. 10 mMTris-Hcl, 100mM NaCl pH 7.4 2. adding CL/DMPC liposomers into Ksv.1.8 mol/l. BPP/lipid 1:50 3.adding CL/ DMPC liposomes into Ksv.1.4 mol/l. BPP/lipid 1:100.

Blue shift of the peak of fluorescence spectrum showed that the BPP environment became more hydrophobic after its interaction with the lipid (Fig. 2), indicating that part of the polypeptide was inserted into the lipid bilayer. In order to investigate further the insertion, fluorescence quenching experiments were used. Quenching constant Ksv was obtained with NaI as a soluble quencher according to the Stern-Volmer equation: Io/I=1+Ksv[Q]. It was shown (Fig.3) that Ksv of BPP solution (3.0 mol/l) was higher than that after its interaction with lipid (1.8 mol/l for BPP/lipid: 1:50 and 1.4 mol/l for BPP/lipid : 1:100), demonstrating penetration of BPP into the membrane. The depth of insertion was studied with spin-labeled PC, 5-DOXYL PC and12-DOXYL PC containing liposomes according to lit. [4]. Zcf, the distance between fluorescent amino acid and the center of the lipid bilayer was calculated to be 136 nm, indicating an insertion depth of about 14 nm below the head group of phospholipids.

The secondary structure was studied by FT-IR [5].The results of deconvoluted spectra and curve-fitted amino I bands of BPP showed that α helix content increased from 10% to 20%, random coil decreased from 48% to 35% before and after interaction with lipid (Fig.4, Fig.5).

Acknowledgment

We acknowledge the financial support from the National Nature Science Foundation of China, and Dr. R.E.Chance, Eli Lilly Comp, USA who sent us BPP.

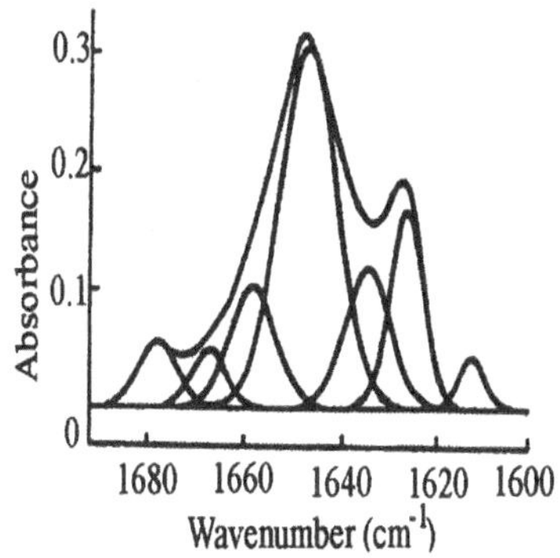

Fig. 4. Deconvoluted spectrum and the curve-fitted amide I' bands of BPP.

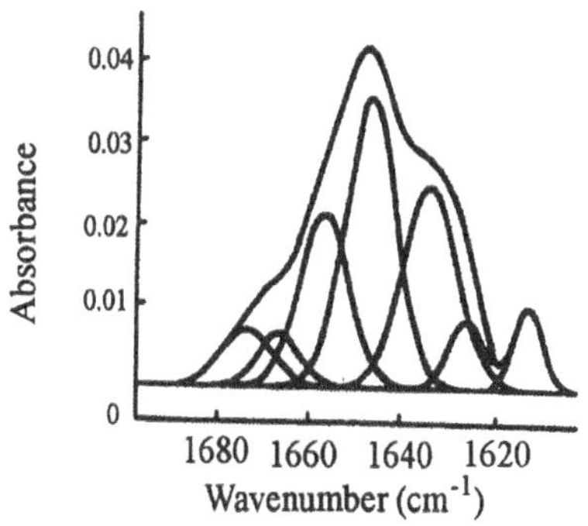

Fig. 5. The curve fitting to the deconvoluted spectrum of BPP after interaction with CL/PC liposomes.

References

1. Lin, T.M., In George, B.J.G. (Ed.) Gastrointestinal Hormones, Raven Press, 1980, p. 276.
2. Gu, Z.F., Nie, S.Q., Wang, Z.J. and Lin, K. C., Acta Biophysica Sinica, 4 (1988) 179.
3. Yang, L., Nie, S.Q., Chen, H.Y. and Lin, K.C. Acta Biophysica Sinica, 11 (1995) 25.
4. Chattopadhyay, A. and London, E., Biochemistry, 26 (1987) 39.
5. Mantsch, H.H and Surewicz, W.K, In V.Renugopalakrishnan, P.R., Carey, I.C.P., Smith S.-G., Huang, A.C. and Storer (Eds.) Protein, ESCOM Press, The Netherlands, 1991, p. 125.

Preparation and evaluation of microspheres with biodegradable PLGA(50/50) containing an LHRH antagonist

Qi-Ming Xia[a] and Roland Bodmeier[b]

[a]*Tianjin Municipal Research Institute for Family Planning, Tianjin 300191, China*

[b]*College of Pharmacy, The University of Texas at Austin, Austin, TX 78712-1074, U.S.A.*

Introduction

Copolymers of D,L-lactic acid-co-glycolic acid (PLGA) are the widely used biodegradable artificial polymers for sustained-release formulation [1]. There are several LHRH agonists which have been successfully microencapsulated in PLGA by a phase separation technique [2,3]. The LHRH antagonist (TX5) [4], [NAc-D2Nal^{1}, DpClPhe^{2}, DPhe^{3}, Arg^{5}, D3Pal^{6}, DAla^{10}]LHRH has a high biologic potency and water-soluble. The objective of the study was to prepare TX5-loaded PLGA(50/50) microspheres (Ms) by oil in oil solvent evaporation method [5], and to evaluate the encapsulation efficiency, morphology and controlled release of the peptide *in vitro* of the resulting microspheres.

Results and Discussion

1.The parameters of the constant TX5-loading of 2%(W/W) by oil in oil method were as follows:

Amount of PLGA (50/50), mg:	200;
Amount of TX5, mg:	4.10;
Internal organic phase (Acetonitrile), ml:	2
External phase (cottonseed oil), ml:	80
Conc. of surfactant (Span 80 HLB 4.3), % :	0.5(W/V)
Temp. of the oil phase, °C:	22±1
Agitation time for solvent evaporation, min:	60

2. Determination of TX5 content in microspheres: 5mg TX5-loaded microspheres were accurately weighed and dissolved in 1ml methylene chloride. TX5 was extracted with 1 ml 0.1M acetic acid and subjected to gradient HPLC analysis [4].

The LHRH antagonist TX5 can be encapsulated in PLGA (50/50) microspheres with high encapsulation efficiency and high yield. The successful entrapment of peptide within the microspheres was associated with a fast rate of precipitation of the polymer from the organic solvent phase and a low solubility of peptide in the oil phase. The microspheres with TX5-loading can sustain release *in vitro* for up to two months at 37 °C, 80 rpm/min. No indications of peptide degradation were observed after extracting the peptide from the microsphere, and analysis by a stability-specific HPLC-method.

Table 1. Batch-to-batch reproducibility of micrographs TX5-loaded PLGA (50/50)

Batch	Encapsulation efficiency(%)	Microsphere yield(%)
1	77.50	95.32
2	83.50	90.03
3	84.50	89.11
4	80.00	93.44
5	82.00	92.23
6	79.50	94.04
average	81.17	92.36
SD	2.64	± 2.40

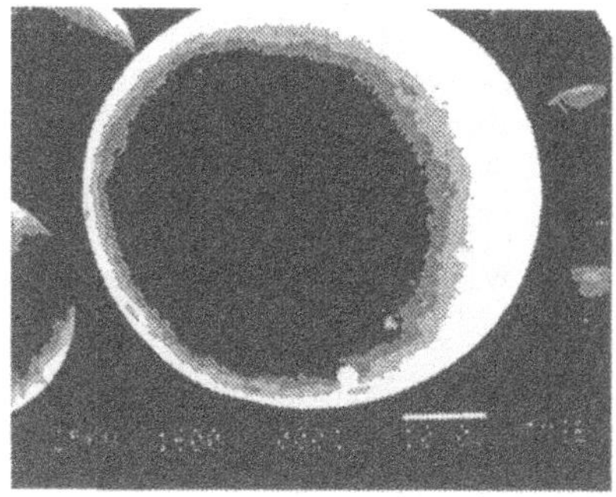

Fig. 1. Scanning electron microscopy (SEM) of the microspheres.

Table 2. Particle size distribution and effect of the particle size on TX5-loading and encapsulation efficiency of PLGA (50/50) microspheres n=6

Particle size(μm)	Weight fraction (%)	Encapsulation efficiency(%)
<45	16.31±4.22	74.50±9.00
45-75	33.42±1.99	80.50±4.50
75-105	23.10±3.35	82.00±9.50
105-150	19.04±4.14	84.50±8.50
150-180	4.82±1.89	83.50±10.50
180-250	2.51±1.03	79.50±9.00
>250	0.80 ±0.10	80.00±9.50

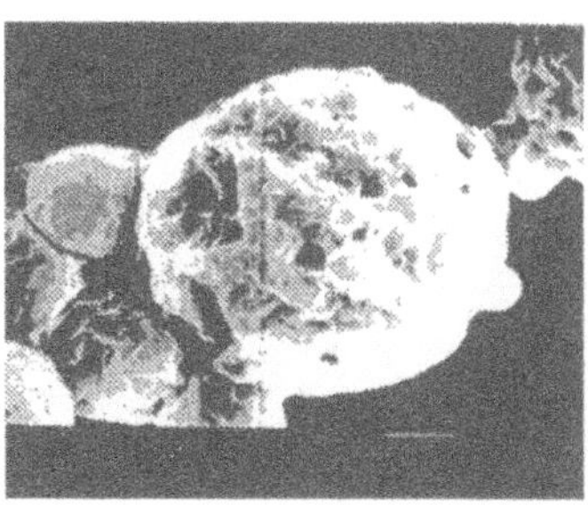

Fig. 2. SEM of the microspheres after releasing, 11 days in double distilled water at 37 °C, 80 rpm/min. at 37 °C, 80 rpm/min.

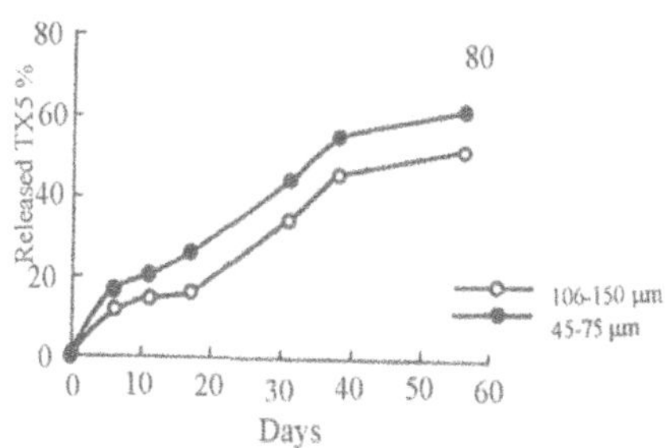

Fig. 3. Release profiles of TX5 from microsphere of different particle in double distilled water.

References

1. Kent, J.S. , Sanders, L.M., Tice, T.R and Lewis, D.H., In Long Acting Contraceptive Delivery System, New York 1984, p.169.
2. Sanders, L.M., Kent, J.S., McRae, G.I., Vickery, B.H., Tice, T.R and Lewis, D.H., J. Pharm. Sci., 73 1984) 294.
3. Ruiz, J.M. and Benoit, J.P., J. Controlled Release, 16 1991) 177.
4. Xia, Q.M., Xiao, S.B., Fang, X., Huang, H., Liu, M.M., Wang, Z.Y. and Wang, J.J., J. of Tianjin Medical College, 13 (1989) 17.
5. Wang, H., MS Dissertation, The college of Pharmacy, The Univ. of Texas at Austin, 1993.

Ester exchange reaction between N-phosphoserine and thymidine

Yuan Ma[a], Xiang-Hong Li[a], Yi Chen[b], Yu-Ping Feng[a] and Yu-Fen Zhao[a]
[a] *Department of Chemistry, Bio-organic Phosphorus Chemistry Laboratory, School of Life Science and Engineering,Tsinghua University, Beijing 100084, China*
[b] *Institute of Chemistry, Chinese Academy of Science,Beijing 100080, China*

Introduction

Proteins and nucleic acids are two of the most important macromolecules in living organisms. Phosphorus plays a significant role in the structure or functions of these two classes of molecules. For example, phosphodiester is the backbone of DNA and RNA, and many proteins' activity were regulated through phosphorylation and de-phosphorylation on their side chains [1] . In the last decade our group had studied the characteristics of N-dialkylphosphoryl amino acids, which embodied the basic architecture of phosphodiester and phosphoproteins in one molecule. It was found that many bio-mimic reactions, such as self-catalysis to yield peptides in organic solvents or even in aqueous solution [2,3], ester exchange reactions on phosphorus in alcoholic media [4,5] could occur on the dialkylphosphoryl amino acids.

Results and Discussion

When the ester is exchanged between the nucleosides and the N-dialkylphosphoryl amino acids, the nucleotides formed [6]. Recently we investigated the interaction between diisopropylphosphoryl serine (DIPP-Ser) and thymidine, which is a deoxynucleoside. With TLC tracing the experiment, it was shown that after prolonged reaction for days in pyridine at room temperature a number of thymidine derivatives were formed. For their structural elucidation, the FAB-MS data of the crude products separated by Sephadex G10 are summarized in Table 1.Through comparing with authentic samples on capillary electrophoresis (CE), 5'-TMP and 5'TpT3' were confirmed as two of the products. Further evidence of the phosphorylated nucleoside was shown by regular and DEPT ^{13}C-NMR spectra. It was shown that there were four peaks around 62-67 ppm for C_5' carbon, of which three were split by phosphorus with $^3J_{P-C}$ = 4.58 Hz, indicating that$_5$C ' was phosphorylated. Simultaneously there were four peaks around 69-73 ppm for C_3' carbon, of which one was also split by phosphorus with $^3J_{P-C}$ = 5.66 Hz.. Thus there might be 3'-TMP and 3'TpT3' or 5'TpT5', as well. Among the several kinds of amino acids tested, except DIPP-Ser, dibutylphosphoryl serine (DBP-Ser) and diisopropylphosphoryl serine (DIPP-Asp) also could have similar reaction.

In general, both nucleoside and deoxynucleoside could yield a nucleotide or deoxynucleotide dimer of nucleic acid in participation of dialkylphosphoamino acids. Based on the above results, it could be presumed that in the origin and evolution of life, phosphoryl amino acid might have played an important role in peptides and oligonucleic acids pre-biotic synthesis.

Table 1. Positive ion FAB-MS and CE data of mixed products of DIPP-Ser with thymidine

Possible products	FAB-MS / M+1 (relative peak intensity)	CE retention time (authentic sample)
O T-P-OH(TMP) OH	326(20)	17.75(17.73)
O T-P-O-isoPr OH	367(80)	
TpT	550(10)	10.62,10.76(10.28)

Acknowledgments

The project was supported by the National Natural Science Foundation of China, the Chinese National Education Ministry and the National Science and Technology Committee and Foundation of Tsinghua University.

References

1. Cohen P., Annual Review of Biochemistry, 58 (1989) 453.
2. Li, Y. M., Yin, Y. W. and Zhao, Y. F., Int. J. Peptide & Protein Res., 39 (1992) 375.
3. Zhao, Y. F., Ju and Y., Li, Y. M., ibid., 46 (1995) 518.
4. Xue, C. B., Yin, Y. W. and Zhao, Y. F., Tetrahedron Lett., 29 (1988) 1145.
5. Tan, B. and Zhao, Y. F., Chinese J. Org. Chem., 15 (1995) 30.
6. Ju, Y., Zhou, W. H. and Zhao, Y. F., Chin. Chem. Lett., 6 (1995) 209.

Cleavage of DNA by N-phosphoserine in histidine aqueous medium

Yuan Ma, Xian-Hong Li and Yu-Fen Zhao
Bio-organic Phosphorus Chemistry Laboratory,
School of Life Science and Engineering,
Tsinghua University, Beijing 100084, China

Introduction

Chemical cleavage of DNA is of considerable importance for both molecular biological study and clinical applications. The commonly used reagents for DNA chemical cleavage are chelates of metal ions, or antibiotics and their mimics, such as Fe(II)-EDTA [1], Cu(I)-bis-(1,10-phenanthroline) [2], Cu(II)-thiols [3,4], dynemicin A [5], calicheamicin γ_1[6] and so on. These systems all require activation by either a strong oxidant or high concentrations of reducing agent to generate oxygen or hydroxyl radicals, which then induce DNA strand scission. Recently we reported N-(O,O-diisopropyl)phosphoryl histidine (DIPP-His) enables to cleavage of DNA without the participation of metal ions [7]. In this paper, we describe a more efficient DNA chemical cleavage system, N-(O,O-diisopropyl)phosphoryl serine (DIPP-Ser) in saturated histidine aqueous medium.

Results and Discussion

DIPP-Ser was dissolved in a saturated histidine aqueous medium, pH 7.0 and stored at -20°C for 20 days. Then the solution was applied as the DNA cleavage reagent. It was found that it caused DNA strand scission in a concentration-dependent manner, thus converting the supercoiled plasmid to the linear form. As the incubation time was prolonged to 48 hrs, DNA incubated with DIPP-Ser was totally transferred into the linear form. In a fresh solution of DIPP-Ser in saturated histidine aqueous medium, however no cleavage was observed. DIPP-Ser in Tris-HCl buffer (pH 7.5) or in imidazole aqueous solution (0.25 M, pH 7.5) also show no obvious DNA cleavage. The storage solutions of DIPP-Ala or DIPP-SerOMe in saturated histidine also showed no cleavage function.

Since incubation in saturated histidine for 24 hrs had only a barely perceptable cleavage effect on DNA, and DIPP-Ser in Tris-HCl buffer or in imidazole had no effect, it seems that neither histidine nor DIPP-Ser alone can effect cleavage on supercoiled DNA. In addition, the fresh solution of DIPP-Ser in histidine also showed no cutting function on DNA. It could be deduced that both histidine and DIPP-Ser together were essential, as well as aging and prolonged incubation in the solution to produce the ability to cleave DNA.

Could it be some new compound formed in the aged mixture of DIPP-Ser with histidine was responsible for the DNA cleavage? Since DIPP-His could cleave DNA, we first considered whether the mixture yielded DIPP-His. The negative ion fast atom bombardment mass spectrometry of the storage solution showed no obvious DIPP-His peak (M-1=318). However, there was an intensed DIPP-Ser-His dipeptide peak (M-1=405) in the MS spectrum. The results were the same as in our previous work that demonstrated

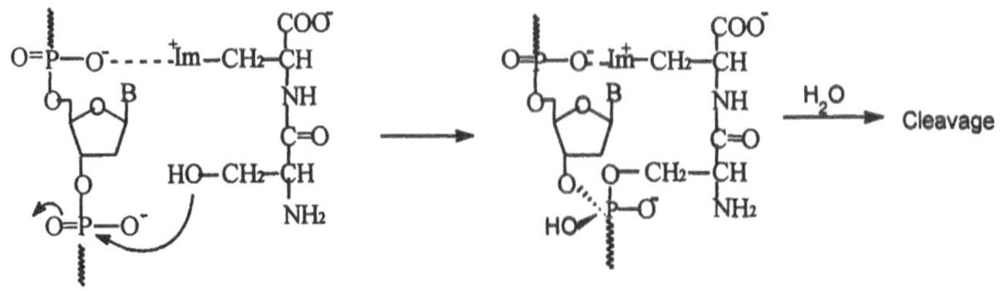

Scheme 1. Presumed mechanism of DNA cleavage by Ser-His dipeptide.

DIPP-amino acids could create phosphopeptide in aqueous medium solution with other amino acids [8]. After loss of the phosphoryl group, N-phospho-Ser-His dipeptide will produce Ser-His dipeptide. It was thought that the Ser-His dipeptide might evince the DNA cleavage phenomenon. After treatment with the authentic Ser-His dipeptide, DNA cleaving was indeed observed. The cleavage mechanism was postulated as Scheme 1. There are simultaneously both a binding site and catalysis site in one molecule. The imidazole group fixed the phosphoryl group through ionic interaction to hold DNA and the peptide together, then a pentacoordinate phosphorus transition state was formed, which was easily hydrolyzed by water to give DNA fragments. Further studies are now in progress to confirm the mechanism.

Acknowledgments

We thank the National Natural Science Foundation of China and the National Science and Technology Committee of China and Foundation of Tsinghua University for their financial support to this work.

References

1. Hertzberg, R.P. and Dervan, P.B., Biochemistry, 23 (1984) 3934.
2. Spassky, A. and Sigma, D.S., Biochemistry, 24 (1985) 8050.
3. Reed, C.J. and Douglas, K.T., Biochem. Biophys. Res. Commun., 162 (1989) 1111.
4. Reed, C.J. and Douglas, K.T., Biochem. Biophys. Res. Commun., 275 (1991) 601.
5. Nicolaou, K.C., Hong, Y.P., Torisawa, Y., Tsay, S.C. and Dai, W.M., J. Am. Chem. Soc., 113 (1991) 9878.
6. Lee, M.D., Dunne, T.S., Chang, C.C., Morton, G.O. and Borders, D.B., J. Am. Chem. Soc., 109 (1987) 464.
7. Li, Y.F., Sha, Y.W., Ma, Y. and Zhao, Y.F., Biochem. Biophys. Res. Commun., 213 (1995) 875.
8. Ju, Y., Zhao, Y.F., Sha, Y.W. and Tan, B., Phosphorus, Sulfur and Silicon, 101 (1995) 117.

RNA cleavage by N-phosphoryl serine in histidine buffer

Xiang-Hong Li[a], Yuan Ma[a], Yi Chen[b] and Yu-Fen Zhao[a]
[a]*Bio-organic Phosphorus Chemistry Laboratory, Department of Chemistry, Tsinghua University, Beijing 100084, China*
[b]*Institute of Chemistry, Chinese Academy of Sciences, Beijing 100080, China*

Introduction

RNA cleavage agents have received considerable attention for multiple purposes, such as the strategy of coupling a RNA hydrolytic catalyst to antisense DNA for site-directed cleavage of RNA in gene therapy [1]. Ribozymes have been described as possible agents against HIV. But low efficiency and high sensitivity to RNases limit their usage. In addition, studies on RNA structure and function have been hampered by the limited kinds of site-specific RNases isolated from living cells [2]. Therefore, it is necessary to develop new strategies for RNA hydrolyzing reagents. The cleavage of RNA via hydrolysis of the phospho-diester backbone has attracted many attention due to advantages over oxidative cleavage [3,4]. It has been shown that N-(O,O-diisopropyl) phosphoryl amino acids possessed many bio-mimic characteristics. For example, they can self-catalyze to form peptides and can conduct ester exchange on phosphorus. This paper describes the RNA cleavage function of N-(O,O-diisopropyl)phosphoryl serine (DIPP-Ser) in histidine buffer and discusses the possible mechanism.

Results and Discussion

Fig. 1 shows the result of agarose gel electrophoresis of yeast RNA incubated with increasing concentrations of DIPP-Ser in saturated L-histidine buffer. The result was confirmed by capillary electrophoresis. The cleavage reaction seems to be specific for DIPP-Ser in histidine buffer, because no such phenomena occurred when Tris Cl (50 mM) or imidazole (0.4 mM) was used instead of histidine as buffer. This phenomena also dis not occur when DIPP-Ala (DIPP-alanine), DIPP-SerOMe (DIPP-serine methyl ester) or serine (20mM) without any modification was used as cleavage agent in saturated histidine buffer.

The above results indicated that DIPP-Ser in saturated L-histidine aqueous solution can cleave RNA at micromolar concentration at neutral pH and 37°C. As the concentrations of DIPP-Ser increased, the extent of the cleavage increased significantly. The cleavage reaction needs both DIPP-Ser and histidine simultaneously. In our experiment, we found that histidine buffer itself also catalyzed the cleavage of RNA. This result was reasonable. It is well known that the imidazole ring in the histidine side chain plays an important role in many enzymes, such as in ribonuclease A where the two imidazole rings of His-12 and His-119 act as both the acid and base catalysts in RNA cleavage. We propose that the mechanism for the cleavage of RNA by histidine is similar to ribonuclease A. What is most interesting is that the extent of cleavage increased significantly after DIPP-Ser was added. Moreover, cleavage occured in a DIPP-Ser concentration dependent manner. DIPP-Ser must play an important role in the cleavage reaction. However, DIPP-Ser without histidine

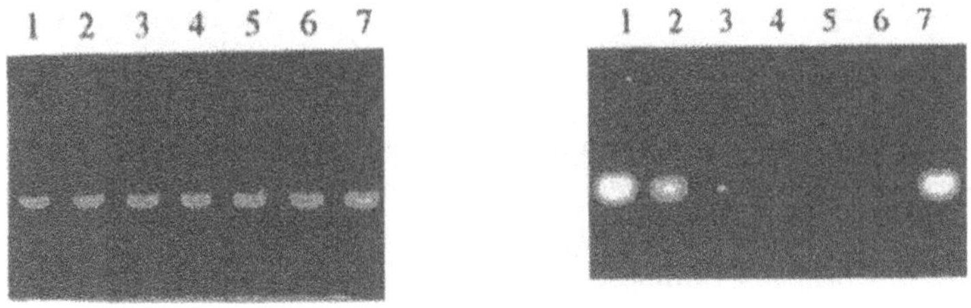

a. Before incubation b. Incubated at 37 °C for 17.5h

Fig. 1. Representative cleavage patterns of yeast RNA by DIPP-Ser and histidine on 1.2 agarose gel. Lanes 2-7, RNA (1 mM in base) was incubated with DIPP-Ser in saturated L-histidine buffer, pH 7.5 at 37 °C for 17.5 hr. Lane 1, the same composition with lane 7, kept at 20 °C. The concentrations of DIPP-Ser from lanes 1-7 are 0 mM, 0.5 mM, 1.0 mM, 1.5 mM, 2.0 mM, 2.5 mM, 0 mM.

had no effect on RNA. Traced by FAB-MS, DIPP-Ser-His (N-O,O-diisopropylphosphoryl Ser-His) dipeptide was found in the DIPP-Ser and histidine mixture. We therefore propose that DIPP-Ser-His was responsible for the cleavage reaction.

Acknowledgments

We thank the National Natural Science Foundation of China and the National Science and Technology Committee of China for their financial support to this work.

References

1. Stein, C. A. and Cohen, J. S. Cancer Res., 48 (1988) 2659.
2. Cedergern, R. and Grosjean, H., Biochem. Cell Biol., 65 (1987) 677.
3. Morrow, J. R., Buttrey, L. A., Shelton, V. M. and Berback, K. A., J. Am. Chem. Soc., 114 (1992) 1903.
4. Modak, A. S., Gard, J. K., Merriman, M. C., Winkler, K. A., Bashklin, J. K and Stern, M. K., J. Am. Chem. Soc., 113 (1991), 283.
5. Li, Y. M., Yin, Y. W. and Zhao, Y. F., Int. J. Peptide and Protein Res., 39 (1992) 375.
6. Zhao, Y. F., Ju, Y., Li, Y. M., Wang, Q., Yin, Y. W. and Tan, B., Int. J. Peptide and Protein Res., 45 (1945) 514.

A world-wide web server of protein domain assignment

Jing-Chu Luo[a], Suhail A. Islam[b], Rob. B. Russell[b] and Michael J.E. Sternberg[b]
[a]*National Laboratory of Protein Engineering and Plant Genetic Engineering and College of Life Sciences, Peking University, Beijing 100871, China*
[b]*Imperial Cancer Research Fund, London WC2 3PX, U.K.*

Introduction

Protein domain is an important structural unit in protein molecules. Identification and assignment of domains from the three-dimensional structure can be accomplished by visual inspection and computer programs. A database of protein domain assignment is useful for molecular biologists. The rapid development of computer networks and the information highway provides means for constructing molecular bioinformatics servers. A World-Wide Web Server of protein domain assignment has been set up on the Internet.

Materials and Methods

The table of protein domain assignment resulting from both literature paper survey and computer program output was used to construct the database of domain assignment for 284 non-redundant protein chains. The database contains various information on domain arrangement, folding type, ellipticity and surface area, as well as PDB code, chain ID and protein name. It can be searched on PDB code or by a keyword of protein name. Interactive assignment of domains based on the 3D coordinates on local machines is available by running the domain assignment program on the server. The coordinates of determined structures were obtained from the Brookhaven Protein Data Bank. Remote users can provide their own set of coordinates to run the program and to get results directly. Various new techniques in computer and information science, such as HTML, Email and network programming were implemented in this work. The HTTP addresses are as follows:
http://bonsai.lif.icnet.uk/domains/domain.html
http://www.lsc.pku.edu.cn/domains/domain.html.

Results and Discussion

A typical output of domain assignment gives some general information. An example of the domain assignment for the Fc fragment of Immuno-globulin is as follows:

```
PDB Code: 1fcl
Chain ID: A
Protein Name: Fc Fragment
Number of residues:   207
X-ray Resolution:     2.9 angstroms
X-ray R factor:       .22
Percentage of Alpha: 12
Percentage of Beta:  51
Surface Area:         13110 angstroms
Ellipticity:          2.9
```

Result of domain assignment based on literature paper and from the program are as follows:

```
Number of Domains: 2
Domain 1
  Number of fragments: 1
    Start from Residue No: 238 (PDB file residue ID)
    Stop  at Residue No:   341 (PDB file residue ID)
  Percentage of Alpha:  13
  Percentage of Beta:   48
  Type of fold:         Beta/Beta
  Number of Residues:   104
  Surface Area:         6729
  Loss of surface Area: 500 angstroms
  Ellipticity:          2.4
Domain 2
  Number of fragments: 1
    Start from Residue No: 342 (PDB file residue ID)
    Stop  at Residue No:   443 (PDB file residue ID)
  Percentage of Alpha:  11
  Percentage of Beta:   55
  Type of fold:         Beta/Beta
  Number of Residues:   102
  Surface Area:         6260
  Loss of surface Area: 508 angstroms
  Ellipticity:          2.6

Domain 1
    Start from residue No: 238 (PDB file residue ID)
    Stop at residue No:    340 (PDB file residue ID)
Domain 2
    Start from residue No: 341 (PDB file residue ID)
    Stop at residue No:    444 (PDB file residue ID)
```

Further development of this domain assignment server, such as expanding the database, improvement of the automatic assignment and inclusion of image presentation, is underway.

Acknowledgment

This work was supported by a joint-research project from the Royal Society, U.K.

References

1. Islam, S.A., Luo, J.C. and Sternberg, M.J.E., Protein Engineering, 8 (1995) 513.

Morphological characterization of peptide vesicle of native protein from Chinese wheat by electron microscopy

Su-Yun Wang and Hong-Lan Fu
College of Life Sciences, Peking University, Beijing 100871, China

Introduction

It is well known that synthetic amphiphilic amino acid mimics can form vesicles or liposomes under certain conditions. However, construction of peptide vesicles from native proteins is a new attractive subject and has some potential for medical application. Various methods of cellular and molecular biology have been used to investigate the combination and insertion of vesicle walls. This paper presents the morphological characterization of peptide vesicles by means of electron microscopy.

Materials and Methods

The peptide vesicle was made from a 35kD storage protein purified from the flour of Chinese wheat *Triticum aestivum*[1]. The method of ultrathin sectioning was used in specimen preparation. After adding agar (1.6%) into the peptide vesicle solution, the specimen was fixed with glutaraldehyde (2.5%) and osmic acid (1%), dehydrated with acetone, embedded in Epon 812. It was then sectioned, stained with uranyl acetate and lead citrate, and examined under a JEM-100CX transmission electron microscope.

Results and Discussion

The diameter of the vesicles was shown to vary between 0.2μm to 2.0μm with a wall

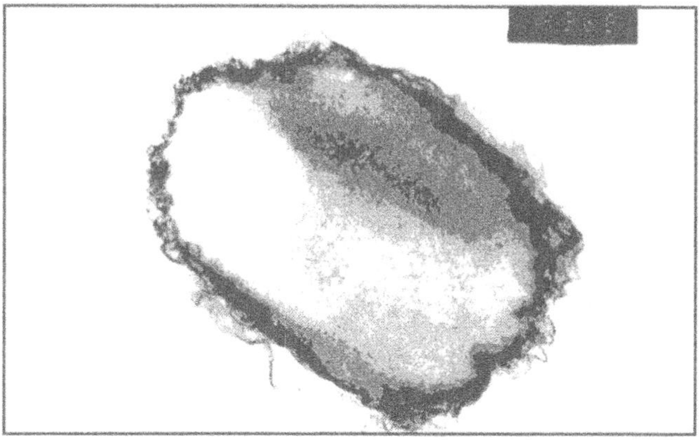

Fig. 1. Ultra thin section of peptide vesicles, about 1.5 × 1.0μm. The wall of the peptide vesicle was attached with fibrils. (X42900)

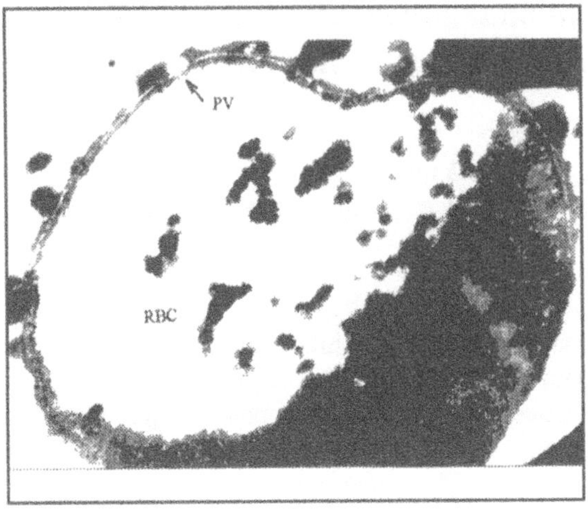

Fig. 2. The peptide vesicles fused with rabbit red blood cell .The bilayer of the red blood cell has changed into a single layer at the fusion position. (X22100)

thickness of 0.02-0.20μm. Fibril attachment can be found at the surface of the vesicles (Fig. 1). Encasement of Even's blue (0.4kD) by peptide vesicles was shown in our experiment. Enclosing a 150kD fluorescent antibody will be carried out to explore its potential medical application as a drug carrier. Dehydration and rehydration can be found under both electron and light microscope. The peptide vesicles can be fused into erythrocyte membrane by constituting the vesicles with membrane fragments of rabbit red blood cells.

References

1. Wang, S.Y. and Yang, Z.H., Acta Scientiarum Naturalium, Universitatis Pekinensis, 32 (1996) 108.

Expression of mouse metallothionein-I cDNA in *E.coli*

Yi Xiong, Yan-Jiao Zhou and Bing-Gen Ru
National Laboratory of Protein Engineering, College of Life Sciences, Peking University, Beijing 100871, China

Introduction

Metallothioneins (MT) are a family of cysteine-rich, low molecular weight, metal-binding proteins [1]. They play an important role in the homeostasis of essential metal ions, e.g., zinc and copper [2], as well as in the detoxification of heavy metals such as cadimum and mercury. They are also important in mammalian UV response. MTs are usually single-chain proteins, without a α-helix or β-sheet, and bind a total of 7 equivalents of bivalent metal ions. In this regard, MTs are ideal proteins for the examination of structure/function relationships by site-directed mutagenesis or other molecular biological methods. The expression of MT in *E.coli* has been reported previously, but attempts to produce high levels of recombinant proteins had only limited success because of their low stability [3,4]. We describe here the expression in *E.coli* of mouse MT-I cDNA as a carboxyl-terminal extension of glutathione-S-transferase.

Results

Two oligonucleotide primers were synthesized and plasmid pBX/MT, containing mMT-I cDNA, was used as template for the polymerase chain reaction. Amplification products of the predicted size (205 bp) were verified on 8% PAGE. These products were then inserted into glutathione-S-transferase(GST) fusion expression vector, pGEX-4T-1 (Pharmacia), to produce plasmid pGMA5 (Fig.1).

E.coli BL21 cells transformed with pGMA8 plasmids were cultured and amplified as described before [5] and the fusion proteins were expressed after $CdCl_2$ and isopropyl-β-D-thiogalactoside (IPTG) were added. The GST-MT fusion protein (33KD) was detected by SDS-PAGE and was about 51% of total crude lysates. The fusion protein was purified by single-step affinity chromatography on glutathione-Sepharose4B (Pharma-cia) [5]and digested with thrombin on the column at 25^{0}C. SpMT was cleaved from GST and purified by gel filtration on Sephadex G50 (Fig.2) and lyophilized.

The contents of bound Cd and thiol group of GST, GST-MT fusion protein and purified SpMT were determined (Table1) by atomic absorption spectro-photometry and Ellman's reagent, respectively. Amino acid analysis of SpMT was performed and gave the expected results.

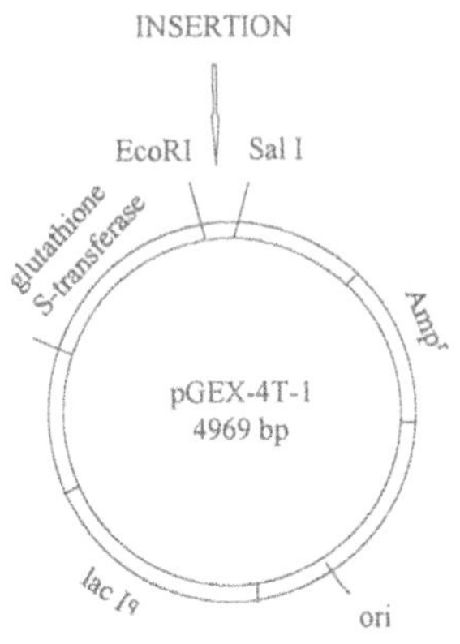

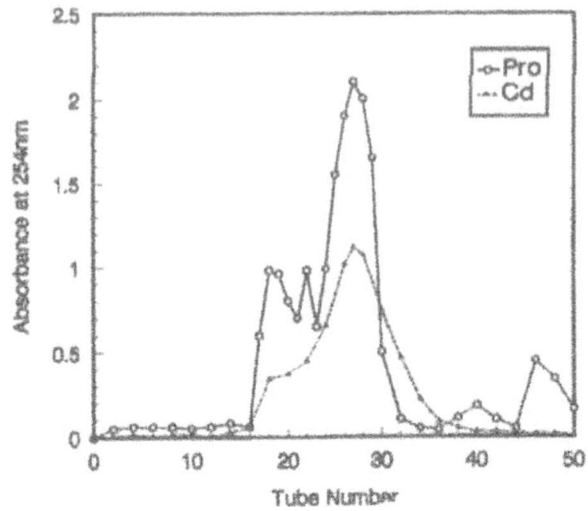

Fig. 1. Construction of plasmid pGMA5.

Fig. 2. Gel filtration pattern on Sephadex-G50.

Table 1. Metal content and thiol content of GST, GST-MT and SpMT

Protein	Cd: Protein	Thiol: Protein
GST	0.14: 1	1.8: 1
GST-MT	6.55: 1	18.5: 1
Sp-MT	6.50: 1	17.4: 1

Discussion

The expression levels of MT in *E.coli* are always limited because of its high content of thiol groups, which are toxic to the host cell. In this paper, a fusion expression vector, pGEX-4T-1, was selected and a much higher yield was attained than previously. The GST-MT fusion protein synthesized in *E.coli* represents approx.51% of total cellular protein, in which the MT portion is about 1/6 8.5%. This is considerably higher than the expression level (0.5%) achieved for mouse [4] or monkey [3] MT in *E.coli*.

Protein content was estimated using a Coomassie blue-based reagent (BIO-RAD) and bovine serum albumin as standard. Both ratios of Cd/protein and thiol/protein are lower than expected, 7:1 and 20:1, respectively, because the reactivities of GST, MT and BSA with Coomassie blue in the protein assay are different. Also the presence of contaminants in proteins will reduce the apparent ratios.

References

1. Kagi. J.H.R. and Schaffer, A., Biochemistry, 27 (1988) 8509.

2. Hamer, D.H., Ann. Rev. Biochem., 55 (1986) 913.
3. Muraoka,Y.M. and Nagaoka,T., Appl. Environ.Microbiol., 53 (1987) 204.
4. Hou, Y.M., Kim, R. and Kim, S.H., Biochem.Biophys.Acta., 951 (1988) 230.
5. Smith, D.B. and Johnson, K.X., Gene, 67 (1988) 31.

Non-radioactive determination of phosphoamino acids in peptides and proteins by capillary electrophoresis

Fa-Yun Che, Xiao-Xia Shao, Lai-Gen Xu, Rong Zeng and Qi-Chang Xia
Shanghai Institute of Biochemistry, Chinese Academy of Sciences, Shanghai 200031, China

Introduction

A new method is described that allows non-radioactive determination of all phosphoamino acids in peptides or proteins. It involves partial hydrolysis of peptides or proteins using 6 N HCl at 110^{0}C for 1.5 hour, then derivatization of the liberated amino acids with phenylisothiocyanate and finally separation of all PTH-phosphoamino acids from other PTH-amino acids or peptides by capillary electrophoresis.The detection limits are 170 fmol for phosphotyrosine, 350 fmol for phosphothreonine and 700 fmol for phosphoserine.

Results and Discussion

Figure 1 shows the separation of all the phosphoamino acid PTH-derivatives. One of the advantages of the method is that there is almost no interference from other amino acids under the analytical conditions. In the electrophoretic condition of pH 2.5, all PTH-phosphoamino acids carry negative charges, whereas other PTH-amino acids carry positive charges or no charges. During electrophoresis, only PTH-phosphoamino acids migrate to the detector and all other PTH-amino acids migrate in the opposite direction and can not be detected. Phosphate electrophoretic buffer can inhibit the hydrolysis and the β-elimination of phosphoester bonds of PTH-phosphoamino acids. The analytical conditions are: fused silica capillary, 72 cm long (50 cm to detector), 50 μm I. D.; buffer: 25 mM sodium phosphate, pH 2.5, containing 15 mM sodium chloride; voltage: -22 kV ; absorbance of PTH-derivatives monitored at 261 nm.

Kemptide (L R R A S L G) was phosphorylated in a reaction catalyzed by the catalytic subunit of protein kinase A. A phosphothreonine-containing peptide (K R T-P L R R) was obtained by incubating the native peptide with protein kinase C and ATP. The phosphopeptides were purified by RP-HPLC. We applied the method to analyze the phosphorylation sites in the above two phosphopeptides and a phosphotyrosine-containing peptide (E D N E Y-P T A R Q G A) (Fig. 2). Two natural phosphorylated proteins, β-casein and phosvitin, were also analyzed (Fig. 2). A higher concentration of trifluoroacetic acid in the sample solution can destroy PTH-phosphoamino acids during electrophoresis. 0.1% or 0.2% trifluoroacetic acid in the sample solution is recommended for the analysis of PTH-phosphoserine.

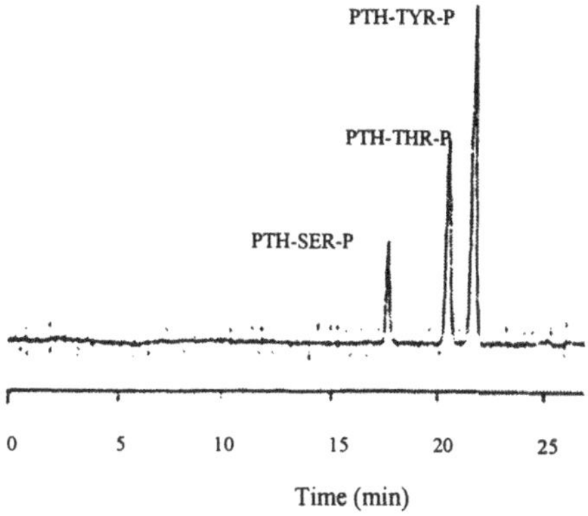

Fig. 1. Separation and identification of PTH-phosphoamino acids. 2 nmol phosphoserine, phosphothreonine and phosphotyrosine, each derivatized with phenylisothiocyanate. PTH-derivatives dissolved in 20 μl sample solution (0.2 % TFA, 20 % CH_3CN, 25 mM NaCl) and 7nl injected into the capillary.

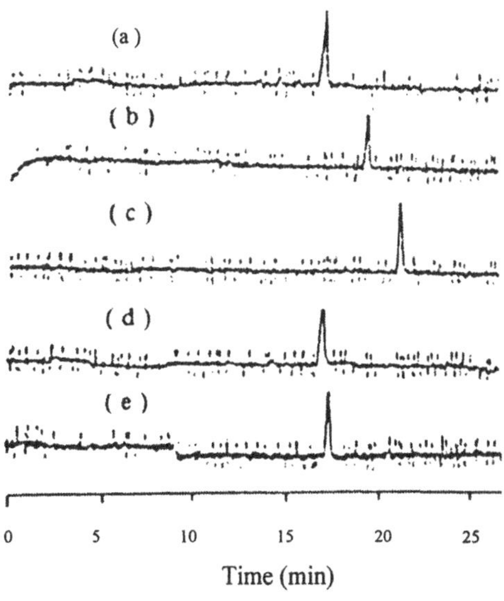

Fig. 2. Identification of phosphoamino acids in peptides or proteins. Samples were partially hydrolyzed and derivatized with PITC. PTH-derivatives were dissolved in 5 μl sample solution except for the analysis of phosvitin, in which PTH-derivatives were dissolved in 100 μl sample solution; (a) 2 nmol L R R A S-P L G, injection 5 sec; (b) 1 nmol K R T-P L R R, injection 4 sec; (c) 1 nmol E D N E Y-P T A R Q G A, injection 1 sec; (d) 1nmol β-casein, injection 2 sec; (e) 0.5 nmol phosvitin, injection 2 sec.

References

1. Meyer, H.E., Hoffmann-Posorske, E., Donella-Deana, A. and Korte, H., In Hunter, T. and Sefton, B.M. (eds.) Methods in Enzymology, Vol. 201, Academic Press, New York, 1991, p. 206.
2. Heber, M., Liedtke, C., Korte, H., Hoffmann-Posorske, E., Donella-Deana, A., Pinna, L.A., Perich, J. ,Kitas, E., Johns, R. B. and Meyer, H. E., Chromatographia, 33 (1992) 347.

Purification of a 28kD protein from maize pollen and studies on its properties

Jian-Hua Wang[a], Ying-Jie Wu[b] and Xian-Rong Wu[b]
[a]*Laboratory of Protein Engineering, Institute of Biotechnology, Beijing 100071, China*
[b]*College of Biological Sciences, Chinese Agricultural University, Beijing 100094, China*

Introduction

A 28kD protein with ATPase activity was purified from the cytoplasmic extracts of maize pollen. It was shown that the 28kD protein was obviously different from the known motor proteins such as kinesin or dynamin. It was probably a new microtubule-associated protein with low molecular weight.

Results and Discussion

A new 28kD protein was purified by ammonium sulfate fractionation, DEAE-Sephadex A_{50} and Mono S ion-exchange chromatography (Fig.1). Its isoelectric point was pH8.3 determined by IEF-PAGE. Western-blotting analysis indicated that the 28kD protein did not have specific immuno-reactions with anti-kinesin monoclonal and anti-dynamin polyclonal antibodies. The maximum ultraviolet absorbance was at 278nm. CD spectrum analysis

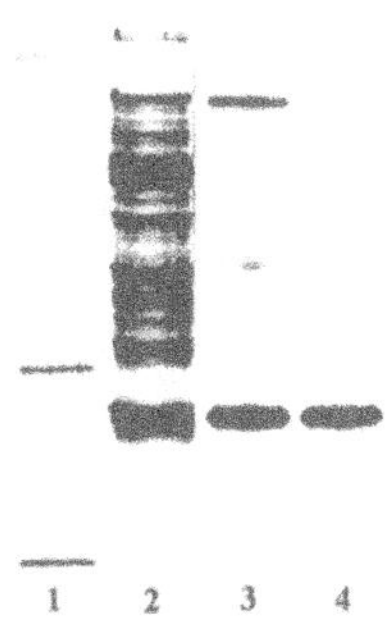

Fig. 1. SDS-PAGE of the 28kD protein. Lane 1, Standard MW markers; Lane 2, Crude extracts; Lane 3, Fractions after DEAESephadex A_{50} column; Lane 4, Preparation purified by Mono Q column.

indicated that the content of α-helix was 26.56%, β-sheet 15.97%, and random coil 51.88%, and the 28kD protein existed in the form of globulin. Pharmacological study showed that its ATPase activity was strongly inhibited by Na_3VO_4 but insensitive to NEM, it was inhibited about 50% by NaF. Oligomycin, KNO_3 and ouabain had no inhibitory effects on the ATPase activity of 28kD protein (Table 1). The above results demonstrated that the 28kD protein is probably a new microtubule- associated protein and may play an important role in the transport of substances along the microtubule in the pollen.

Table 1. Pharmacological Characterization

Inhibitor	ATPase activity (% control)
	100
100μM Na_3VO_4	13
1m MNEM	85
2.4m MNaF	51
100μM ouabain	100
10μg/μ oligomycin	108
100mMKNO_3	96

Acknowledgment

This work was supported by National Laboratory of Biomacromolecules, Institute of Biophysics, Academia Sinica.

References

1. Gelfand VI, Nature ,359 (1992) 480.
2. Valle, R.B. and Shpetner H.S., Annu. Rev. Biochem., 59 (1990) 909.
3. Takehiko, W. and Martin, F., Biochem.Biophys.Res.Comm., 52 (1976) 195.
4. Vale, R.D., Cell ,42 (1985) 39.
5. Shpetner, H.S.and Vallee, R.B., Cell, 59 (1989) 421.

Synthesis of peptides using a thioredoxin gene fusion expression system: An application to hirudin

Xu-Dong Zhu, Du-Sheng Cheng, Xiu-Zhen Li and Feng-Yun Liu
Institute of Biotechnology, Beijing 100071, China

Introduction

Hirudin is a polypeptide of 65 amino acids that acts as a very specific inhibitor of α-thrombin,which plays a central role in blood clotting, because of its intrinsic properties, i.e, strong affinity to α-thrombin, low antigenicity and low toxicity. In this article, we tried to obtain a large amount of hirudin by genetic engineering. Because hirudin produced in *E.coli* has a half-life of only 10-15 min [1], making direct expression of hirudin difficult, the thioredoxin gene fusion expression system was used.

Results and Discussion

The hirudin gene was synthesized using codons typical for highly expressed *E.coli* proteins. An ATG codon for Met at the 5' end was added to allow the cleavage of fusion protein with cyanogen bromide for hirudin recovery. Using the thioredoxin gene fusion expression system [2], the hirudin gene was designed to be expressed as a fusion protein with thioredoxin under the control of the P_L promoter. *E.coli* GI724 cells were transformed and colonies were screened by PCR.The plasmids of three positive colonies were sequenced.

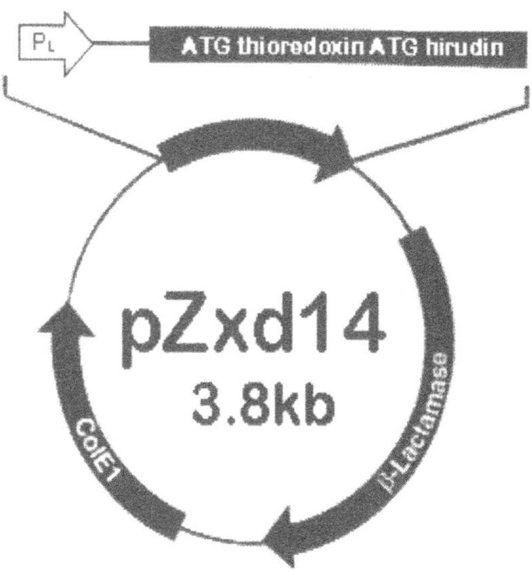

Fig. 1. Structure of pZxd14.

A plasmid with the correct nucleotide sequence was designated pZxd14 (Fig 1). The bacterial cells harboring pZxd14 were induced to express the fusion protein.The cells were osmotically shocked and the fusion protein was released into the shock fluid [2], and reacted in 70% formic acid with cyanogen bromide.The mixture was lyophilized and desalted on Sephadex G-25.The biological activity of the purified product was characterized by anticoagulant assay [3] and colorimetric assay using the thrombin chromogenic substrate Chromozym TH [4] (Data not shown).The yield of hirudin was at least 6mg/liter of bacterial culture.

Acknowledgment

We thank Prof. Hai-Tao Wang for the gift of the thioredoxin fusion expression system.

References

1. Waleh, N.S., Sohel, I. and Lazar, J.B., Gene,117 (1992) 7.
2. LaVallie, E.R., DiBlasio, E.A., Kovacic,S., Grant,K.L., Schendel, P.F. and McCoy, J.M., Biotechnology,11 (1993) 187.
3. Mao, S.J.T., Yates, M.T., Owen, T.J. and Krstenansky, J.L., Biochemistry, 28 (1988) 8170.
4. Chang, J.Y., FEBS Lett., 164 (1983) 307.

Kinetics of self-assembly into oligopeptides from N-phosphoamino acids

Wei-Hua Zhou and Yu-Fen Zhao
Bioorganic Phosphorus Chemistry Laboratory, Tsinghua University, Beijing 100084, China

Introduction

The study of the kinetics of oligopeptide formation by the Fourier transform infrared (FTIR) spectroscopy method [1,2] was reported previously. In this paper, the oligopeptide formation from N-(O,O-dialkyl) phosphoamino acids (DAP-aas) in anhydrous media under low temperature and low concentration conditions were analyzed with biuret reaction. The results indicated that peptide formation followed a first-order kinetics reaction.

Results and Discussion

The peptides derived from DAP-aas were measured using biuret reaction. The relationship between peptide concentration(A) and incubation time(t) was processed as InA vs t and a linear connection was found, suggesting a first-order oligopeptide formation reaction. The rate constants k in different media are shown in Table 1 and the temperature effect on k is listed in Table 2.

Table 1. N-phosphonamino acids and rate constants of peptide formation [$(RO)_2P(O)NHCH(R')COO$]

entry	R group	R' group	in pyridine $k_1(\times10\text{-}5)s^{-1}$	in iPrOH $k2(\times10\text{-}6)s^{-1}$	k_1/k_2
1a	n-butyl	-CH3	1.75	2.62	6.6
1b	i-propyl	-CH3	1.66	2.03	8.2
1c	ethyl	-CH3	1.82	3.10	5.9
2	i-propyl	-OH	1.93	3.74	5.2
3	i-propyl	-COOH	2.04	5.56	3.67
4	i-propyl	$-C_6H_5$	1.50	1.17	12.8
5	i-propyl	$-(CH_2)_4NHP(O)(OR)_2$	1.32	/	/

The peptide formation was proposed to be the result of a series of intramolecular penta-coordinate phosphorus processes of the activated amino group and the activated carboxyl group [3]. Therefore the rate-controlling steps could be assigned to the formation of the intramolecular transition state which is proportional to the concentration of DAP-aas; the expected kinetics should be a first-order reaction. In the presence of nucleophilic side chains such as the hydroxyl and carboxyl groups in DAP-aas, it might be also possible to form a hexa-coordinate transition state which is much more reactive for the activated amino group and carboxyl group formation [4]. Hence it is convincing that in the two

Table 2. The rate content k($\times 10^{-6}$, s^{-1}) and the apparent activation energy(E)

Entry	k in i-PrOH 40°C	55°C	E($kJmol^{-1}$)	k in pyridine 40°C	55°C	E($kJmol^{-1}$)
1b	2.03	9.17	83.8	16.6	65.7	78.3
3	5.56	12.9	117.4	20.4	151	114

reaction systems, DAP-aas with a higher polar side chain will form peptides more easily (Table 1). The substituents in the phosphoryl group also had an observable impact on k, perhaps because of the steric hindrance to the crowded transition state formation. The observed rate constants **1c >1a>1b** suggest that the smaller alkoxyl group is more favourable to the transition state. As for the solvent effect, pyridine might act not only as the base to make the carboxyl group of DAP-aa in the form of anion, but also as the nucleophilic reagent to promote penta-coordinate transition state formation. Thus the apparent activation energy is lower compared to the condition with isopropanol as the solvent.

Acknowledgment

The authors thank the financial support from the China National Natural Science Foundation.

References

1. Li,Y.M., Yin,Y.W. and Zhao, Y.F., Int. J. Peptide Protein Res., 39 (1992) 375.
2. Hu, X.R., Mao,T.Q., Jiang , X.Z., Lu, W.Q. and Sun, S.Q., Proceeding of the 8th Conference on Molecular Spectroscopy, Peking University Press, 1994, p. 25.
3. Hu, R., Mao,Y. and Tao, X., In Lau, A., Siebert, F. and Werncke, W. (Eds) Time -Resolved Vibrational Spectroscopy VI, Springer-Verlag, 1993, p. 260.
4. Zhao, Y.F., Li, Y.M. and Yin,Y.W., Science in China (Series B), 36 (1993) 1451.

Inhibition of trypsin by N-phosphoamino acids

Wei-Hua Zhou and Yu-Fen Zhao
Bioorganic Phosphorus Chemistry Laboratory, Tsinghua University, Beijing 100084, China

Introduction

The serine proteinases and their inhibitors play important roles in regulating many biological cellular events [1]. There are numerous natural protein inhibitors including some with low molecular weights [2]. The N-phosphoamino acids, containing the tetrahedral phosphoryl group that resembles the unstable transition state involved in the catalytic reaction of serine proteinase, and the amino acid side chains that may act as natural substrates to serine proteinase, probably inhibit proteinase by the transition-state-analog mechanism proposed by Robertu [3]. The result presented in this paper indicates that N-(O,O-dialkyl) phospho amino acid (DAPaas) are novel trypsin inhibitors.

Results and Discussion

The trypsin, after being incubated with DAPaas at 25°C, gradually lost its esterase activity (Figure1). Different DAPaas showed different inhibition abilities (Figure 2). N-(O,O-dibutyl)phosphoalanine, N-(O,O-diisopropyl) phosphoalanine (DIPP-ala) and DIPP-phe at mM concentrations can inactivate trypsin completely in 20 h. However, DIPP-asn, DIPP-arg, DIPP-ser and DIPP-asp inhibited only 40%~60% of trypsin activity.

The catalytic activity of trypsin is contributed to a triad charge relay net of Ser195, His57 and Asp102 residues [4]. A tetrahedral intermediate(TI) through the nucleophilic attacking of the carbonyl carbon by the ser195 residue is involved in peptide bond hydrolysis. The DAPaas, assuming a transition-state-like tetrahedral structure at the phosphorus center

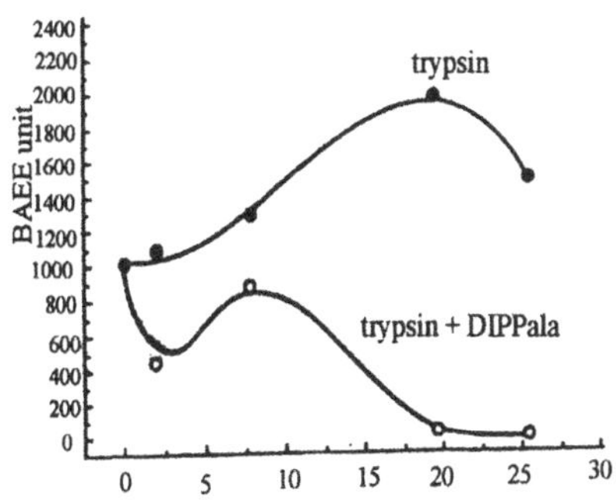

Fig. 1. The esterase activity of trypsin and trypsin+DIPPala.

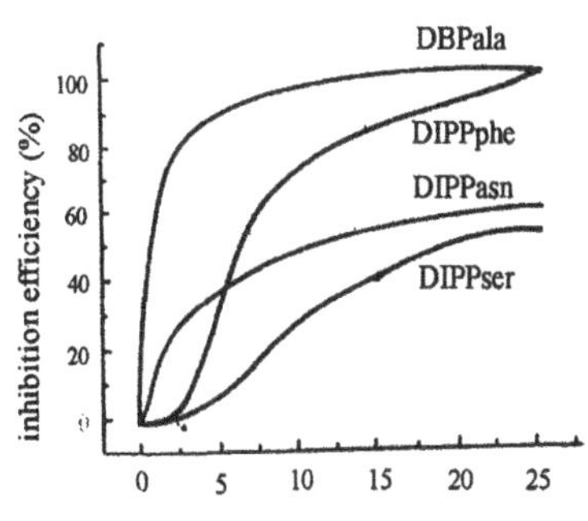

Fig. 2. The inhibition curves of DAPaas.

should fit spatially to the hydrophobic catalytic site and be trapped by the active site. Thus the less polar DAPaas can gain access relatively easily to the hydrophobic active site. The trapped DAPaas, due to their high reactivity, for example, intramolecular phosphoryl group migration from amino group to hydroxyl group [5], react with the active residues resulting from spatially close contact. The ^{31}P NMR signals at -1 ~ -2 ppm characteristic of alkylphosphate for purified trypsin suggested the formation of a serine/threonine adduct in the reaction of DAPaas with trypsin. The result with BSA inferred the formation of adduct only on the special protein residue. In most cases, the nucleophilic ability of the hydroxyl group of Ser195 may be enough to attack the electron-deficient phosphorus in DAPaas to form a stable tetrahedral adduct because the migration from phosphorus nitrogen bond to phosphorus oxygen bond is thermodynamically favoured (ΔG= -30kJ/mol).
DAPaas can inhibit trypsin. The inhibition efficiency varied with molecular polarity. The lower polarity molecules like DAP-ala, DIPP-phe could inactivate trypsin completely. However, the higher polarity molecules like DIPP-ser, and DIPP-asn, only inhibited 40% ~60% of trypsin activity. The transition-state-like phosphoryl group with its high reactivity might be responsible for the inhibition ability.

References

1. Fritz, H., Tschsche, H., Greene, L.J. and Truscheit, E., (eds) Proteinase Inhibitors (Proceedings of the 2nd International Research Conference), New York, Springer-Verlag, 1974, p. 751.
2. Laskowiski, Jr. M. and Kato, I., Ann. Rev. Biochem., 49 (1980) 593.
3. Robertu, J.D., Kraut, J. and Birktoft, R.A., Biochemistry,11 (1972) 4293.
4. Craik, C.S., Argman, C.L., Fletcher, T., Barr, P.J., Fletterick, R. and Rutter, W.J., Science, 228 (1985) 291.
5. Xue, C.B., Yin, Y.W. and Zhao, Y.F., Tetrahedron Lett., (1988) 1145.

Conjugation of galactosamine and ricin

Fen-Wei Wang, Hong-Chun Tang, Ge-E Li, Dun-Hua Dai and Shuo Zheng
The Institute of Pharmaceutical Chemistry of PLA,
Beijing 102205, China

Introduction

Galactosamine is a compound of live affinity, which suggests that it may be used as a carrier for other drugs. In this work, the possibility of using galactosamine as a carrier for protein drugs is investigated. Ricin toxin, chosen as a model protein, has been tested. We successfully linked D-galactosamine hydrochloride with ricin and obtained a purified conjugant.

Method and Results

D-galactosamine hydrochloride was prepared from sodium chondroitin. Sodium chondroitin was dissolved in an appropriate amount of water by stirring at warm temperature, then concentrated hydrochloric acid was added to neutralize the reaction liquid, until it had strong acidity, then stirring and reflux for 2hrs . The completion of the hydrolytic reaction was detected by TLC, after which active carbon was added for decolorization, then filtering and evaporating to remove the water. The residue was purified by recrystallization from the mixed solvent alcohol-acetone. A white crystal is obtained and dried in vacuum. M.P. 178-179^{0}C $[\alpha]$ +90^0. Ricin was purified from caster bean. Galactosamine and ricin were conjugated through their $-NH_2$ and -COOH groups with linking agents N-(3-Dimethylaminopropyl)-N'-ethylcarbodiimidehydrochloride (EDC). Ricin was dissolved in phosphate buffer(pH7.0), then galactosamine was added. After mixing well, EDC was added and dissolved completely, then the pH of the reaction solution was adjusted to 7.0 with 1M$NaCO_3$, and kept at 23^{0}C for 24 h. Purification was performed by Sephadex G-25 column chromatography, the protein eluate was pooled and dialyzed against distilled water, then concentrated and stored at 4^{0}C for use.

The conjugate was identified by the phenol sulfate method. The phenol-sulfate reagent can give a color reaction with the hexose group in the conjugant accompanied by maximum absorption at 490nm. Phenol-sulfate reagent was added to the diluted conjugate solution, ricin solution was also added separately as contrast, and 20min later OD_{490} was detected. As a result, the conjugate reaction mixture appears as an obvious orange color compared to the ricin reaction mixture, proving that galactosamine was successfully conjugated to protein.

References

1. The technique of protein conjugation, The Medical Institute of Army,1989.
2. Zhang, W.J., The Research Method of Complex Multisaccharide, 1987.

Author index

Subject index

The manufacturer's authorised representative in the EU is Springer Nature Customer Service Centre GmbH, Europaplatz 3, 69115 Heidelberg, Germany. If you have any concerns regarding our products, please contact ProductSafety@springernature.com

Printed and bound by CPI Group (UK) Ltd, Croydon, CR0 4YY

29/07/2026

02180607-0013